Mark

wheel-drive instructor and competitor from Hawke's Bay. Mark's upbringing in the South Island's Mackenzie Country was the beginning of a life lived off-road, from the snowy tracks of the Southern Alps to the muddy slopes of his Hawke's Bay hill-country station, Waipari. Find Mark on Facebook at Mark Warren, Waipari Station and Skid Sprint UTV.

MANY A MUDDY MORNING

MARK WARREN

HarperCollins*Publishers*

HarperCollins_Publishers_

First published in 2018
by HarperCollins_Publishers_ (New Zealand) Limited
Unit D1, 63 Apollo Drive, Rosedale, Auckland 0632, New Zealand
harpercollins.co.nz

HarperCollins_Publishers_
Level 13, 201 Elizabeth Street, Sydney NSW 2000, Australia
Unit D1, 63 Apollo Drive, Rosedale, Auckland 0632, New Zealand
A 53, Sector 57, Noida, UP, India
1 London Bridge Street, London, SE1 9GF, United Kingdom
Bay Adelaide Centre, East Tower, 22 Adelaide Street West, 41st floor, Toronto,
    Ontario M5H 4E3, Canada
195 Broadway, New York NY 10007, USA

A catalogue record for this book is available from the
National Library of New Zealand

ISBN: 978 17755 4108 0 (paperback)
ISBN: 978 17754 9139 2 (ebook)

Cover design by HarperCollins Design Studio
Front cover image by Lee Warren, Fotoshoot; tyre tracks by HarboArts.com
Back cover images courtesy of Mark Warren
Cartoons on pp. v, 307 by George Williams
Typeset in Sabon LT by Kirby Jones
Printed and bound in Australia by McPherson's Printing Group
The papers used by HarperCollins in the manufacture of this book are a
natural, recyclable product made from wood grown in sustainable plantation
forests. The fibre source and manufacturing processes meet recognised
international environmental standards, and carry certification.

Dedicated to the unsung heroes of
New Zealand farming

Bit of a cliffhanger (page 169)

CONTENTS

FOREWORD

My involvement as accountant, consultant and listener with Mark Warren, his father and grandfather has spanned 40 years. What are my impressions of Mark, and what have I learned about him?

- Mark has a relatively long memory, but can have a relatively short fuse.
- He can be impatient, doesn't suffer fools gladly and is still learning how to package 'no' effectively.
- He is articulate, can write well, will read if it is important, admits his mistakes and even learns from some of them.
- He will fool and confuse you with his speed and willingness to discuss with you thoroughly a wide range of issues — you will need to press the 'pause' button early on. Once you get behind this veneer, you start to realise that he is well worth listening to, and has the ability to add real value.
- His history behind the farm gate regarding production and financial issues is very sound, and he would be a good

business mentor. Outside the farm gate, he is also sound — his life experiences, and how he has coped with them, are well worth reading.

- He is a very capable farmer and businessman — these are key traits in New Zealand farming today, as these days it's just not good enough to be average. (Maybe this comment now applies in almost all New Zealand businesses.)

- Something I have learned from Mark is that it is better to be proactively wrong rather than to stay with a status quo situation that is just not working — he is a very good example of how being approximately correct is so much better than being precisely wrong.

- Is this book worth reading? Yes, I think you will be surprised at what you will learn — I certainly did, even though I have been hammered with much of it over the years. Who might enjoy this book? Probably anybody who is still breathing.

Pita Alexander
Christchurch
November 2017

A LONG, ROUGH ROAD

As a very young child I was never satisfied with riding my trike on the concrete pathway. I much preferred to ride it through the muddiest puddle I could find; and when the bald rubber wheel skidded in the mud, I improvised a set of tyre chains made out of binder twine to provide enough grip to successfully navigate the boggiest bits. From then on I seem to have retained a fascination with coaxing wheels through slippery situations.

This book is about my life growing up with a fascination for vehicles. It is also about the sometimes slippery, treacherous conditions that prevailed during my years as a high-country shepherd and a Hawke's Bay farmer. There were a few hairy moments along the way. Not that long ago, at a family funeral, I met a friend from my formative years who seemed amazed to see me. 'I'm surprised you're still alive, doing the things you used to do!' she said. Coming from the daughter of a pioneer ski-plane aviator and test pilot, that spoke volumes!

They say if you throw enough mud at the wall, some of it will stick. This book is also about dreams and perseverance as a farmer, through some very challenging days. I hope my story, as well as being entertaining, is of use to anyone in the rural sector going through tough times. We all have a few spills in life (some of us more than others), but it's how you pick yourself up again that really counts!

Mark Warren
Waipari
November 2017

CHAPTER 1

A WOOLLY TALE

Like all quality yarns, this one begins with wool.

My first memories are of lying in my cot cocooned in the lanolin aroma of a greasy, woolly sheepskin, listening to the sound of the squeaky dredge out in Caroline Bay off Timaru. The fleece I was lying on came from the Grampians Station a couple of hundred kilometres inland in the Mackenzie Country: it was gifted to my mother by the owner, the noted painter Esther Hope, when she visited a few months before I was born.

The Grampians features in a best-selling detective story by Ngaio Marsh, one of the few that the great dame set in the land of her birth, New Zealand. In *Died in the Wool*, the victim of a murder is concealed in a bale of wool, the murderer's ingenious plan being to have his victim exported along with the wool, affording him plenty of time to put daylight between himself and the scene of the crime. But the corpse is discovered when a wool buyer recognises the smell of decaying flesh, familiar to him from his time on the battlefield at Flanders.

Coincidentally or not, according to my mother (who should know) I was conceived on the night she and my father attended a 'Welcome Home' dinner for Dame Ngaio Marsh in the Christchurch suburb of Cashmere. I'm also told that on the day I was born the bells of St Mary's were rung for 20-odd minutes. This would have been over the top if I was just another visit from the stork, but I was apparently the first son to be born to a clergyman in Timaru, my father, Martin, having recently been appointed curate of St Mary's Parish. As my Grandfather Alwyn was then the incumbent Bishop of Christchurch, hopes must have been high that I would go on to become the latest member of the family ecclesiastical team.

You can see where they were coming from, because my clerical pedigree is long and distinguished. My great-great-great-grandfather, Henry Williams, was one of the first missionaries to New Zealand, and was responsible for the translation of the Treaty of Waitangi from English into Maori. Throughout the country he was affectionately known as Te Wiremu (for William), and it was said among the Maori that no European had greater mana than Te Wiremu. My other great-great-great-grandfather, William, was involved as well, and went on to become Bishop of Waiapu in Napier, Hawke's Bay. The Williams boys were themselves sons of clerical stock, their father being a Congregational minister in Gosport, back in England. I am directly descended from both Williams brothers, on account of the fact that Henry's son (Archdeacon) Samuel Williams, my great-great-grandfather, married William's daughter Mary. Their daughter Lucy Williams married TJC 'Jack' Warren, a wine merchant who arrived in New Zealand from England via Sydney, bearing a letter of introduction to Hawke's Bay society addressed to Mrs Tom Lowry of Okawa, Hastings, one of Hawke's Bay's noted early settlers. Mrs Lowry invited TJC to attend a woolshed dance at Pourerere (just down the road from where I presently live), where he met Lucy. Taking a shine to her, he went and asked Samuel if he could step out with his daughter.

Perhaps the archdeacon was interested in the fact that TJC had solid connections to a promising source of Bordeaux wine but, to his credit, he didn't just swap his daughter's hand for a few cases of communion wine.

'Who are you?' he is reported to have asked. 'If you mean to walk out with my daughter, young man, you'll jolly well have to prove yourself.'

Undaunted, TJC went to Wellington, where he bought a load of wheat that was going cheap, it having arrived from Sydney wet from a rough Tasman crossing. He had it put in half-sacks, which were then hung out on a fence to dry. He was then able to sell it back to the same merchant who had flogged it to him, but at double the price.

That did the trick. TJC Warren and Lucy Williams were married with the archdeacon's blessing. TJC went on to secure the import licence for a popular brand of sheep dip. This brand was always prominently advertised in the inside cover of Williams and Kettle farm diaries, so there's a good chance Samuel had a quiet word with his cousin Fred, the 'Williams' in the Williams and Kettle stock agency. After a time in Wellington, TJC and Lucy returned to Hawke's Bay, where Samuel built them a large wooden homestead, Penlee House (more recently renamed Langton House), which is still proudly inhabited by my cousin Hugh McBain's family. Doubtless due to his now proven business acumen, TJC was appointed bookkeeper to Samuel's rapidly expanding farming empire. So while I'm something of a disappointment in the Warren family tree — after all, in six generations of that side of the family, it's only TJC and myself who were not directly answerable to God — I like to think I inherited some of TJC's business sense, along with an appreciation of fine wine.

On my mother's side, I am descended from the Scottish clan Sinclair. Her dad, my English grandfather, was Perceval Wallace Jayne, a career Army man who fought in the First World War, later trained at Sandhurst and rose to the rank of lieutenant-colonel. During the Second World War, he was stationed at the RAF base in East Anglia on 'special duties' to His Majesty. He survived the crash of a Wellington bomber in which he was flying.

I never really knew what 'special duties' entailed. Every time we asked, we were shushed. I was always very close to my Aunt Miriam, who lived in British Columbia, and while she trusted me enough to sort out her estate, she would cut conversation short if I asked about her father's war. The fact that he looked almost identical to King George may well have had something to do with it. If he had been assigned to serve as His Majesty's body double, it would have explained why my family was invited to meet Her Majesty, the Queen Mother, when she visited Timaru in 1966. I remember that day reasonably well.

The first inkling I had that something out of the ordinary was afoot was when I was summoned home from primary school around lunchtime. I triked home to where we were living in the vicarage at Geraldine, and found everyone in a state of some excitement. I was then dressed up in my best, my face given a scrub and my hair combed, and we were bundled into the car for the drive to Timaru. There were quite a few well-dressed strangers around — the Queen Mother's entourage of ladies-in-waiting and footmen, as it turned out — and they got into a matching pair of black Austin Sheerlines. We headed south in convoy. As we approached Timaru, we found ourselves driving along between crowds flanking the road and waving Union Jacks, and soon we were caught up in a bit of a traffic jam. There was a roadblock. Dad explained to the policemen manning it that he had an appointment with Her Royal Highness, and a police escort was hurriedly arranged. We swept through Timaru, our little Ford Cortina 1500 station wagon sandwiched between the two gleaming, regal Sheerlines, led by a cop car with its red light flashing. The locals wondered what their vicar had been up to.

We arrived at the quay where the royal yacht *Britannia* was docked. HRH arrived in the royal entourage, having flown in from Te Anau aboard a DC3.

It must have been a rough flight, because when she appeared she looked pretty green about the gills. She came and greeted my mother like an old friend.

9

My sister Celia, who was only three, understood that we were about to board the ship, but she thought we had to pay. She kept trying to press the penny she had fished from the pocket of her coat into the Queen Mum's hand. Eventually (and probably to my parents' relief), it fell through the slats of the gangplank into the harbour.

'You look exhausted, Ma'am,' mum said. 'Forget about us. You go and get a hottie and a cuppa, and get good and ready for the rest of your engagements.'

Her Majesty, the Queen Mother, looked profoundly grateful and accepted.

<hr />

I don't remember much of my first-ever trip to the Mackenzie Country, because I was in utero at the time. Possibly to hurry me up — I was showing no inclination to meet my appointment with the world — Dad loaded Mum into his three-wheeler, two-seater Messerschmitt 'bubble car' and set off to visit a childhood friend, Caro Murray, who farmed at the Wolds Station about 12 kilometres south of Tekapo. In those days it wasn't the tarsealed highway it is now, but a rough shingle track of the sort best tackled in a larger vehicle, and definitely not in a glorified scooter with wheelbarrow tyres.

The rough trip must have worked, because I was born a short time later.

The Grampians Station features in my early memories, too, along with other Mackenzie Country locales. I remember visits with Mum and Mrs Hope, travelling in Mrs Hope's big, old Plymouth. I can't have been more than three, and I remember being carsick on the winding gravel roads. Needless to say, at that age I had no appreciation of how privileged I was to have accompanied Mrs Hope on her painting trips, but I felt lucky enough. Mum and I would go off and amuse ourselves for a few hours while Mrs Hope scrambled up to a vantage point to sketch classic high-country scenes. I remember one such trip to the base of Mount Maggie, beside the Haldon Station woolshed, where Mrs Hope must have forgotten to hand over the keys to the locked car. It was a desperately hot day, and all our water and food was inside the car. Fortunately, the artist mate she had brought along had finished his sketching earlier than she did, and he came to our rescue. He found an ice axe in the boot (Mrs Hope was in the habit of carrying this around and using it to grub any briar rose bushes she came across), and he used it to break into the car to rescue our lunch and water bottle. Afterwards, he showed me how you use an ice axe, cutting steps in the clay bank that we pretended was a perilous, snowy slope. He was a kindly, patient man. It was lost on me at the time that this Sydney Thompson was also one of the country's most eminent painters.

Through Mrs Hope, Mum made many fast and life-long friends amongst the high-country station-owners, and trips to visit them became welcome breaks in the monotony of growing up in the vicarage. I remember staying with Ian and Cecily Innes at Haldon Station next to Lake Benmore in the Mackenzie Country. One day, I discovered in the sunroom at the Haldon homestead a very lifelike but rigid dog that would allow me to sit on his back

11

and pat him for hours, without growling or offloading fleas onto me. Ian Innes explained that the bronze statue was modelled on a sheepdog of his, named Haig after the bottle of scotch he swapped for it. That dog now stands on a stone plinth on the shores of Lake Tekapo next to the Church of the Good Shepherd, casting his keen gaze out to the slopes of Mount John for stragglers. He was placed there as a tribute to the essential musterer's tool of the Mackenzie Country, and features in countless tourist selfies. But how many people know the full story of that dog?

According to my mother, the first recognisable word that I correctly affixed to anything was the word 'truck', as a council tip-truck rolled by. That's a glimpse into what drew my fascination from the very start. If it had four wheels, it warranted my attention, but if it happened to be a Land Rover, it would likely have me drooling. These sturdy, boxy machines were essential tools to the folk of the high country, and I learned at an early age that a ride in a Land Rover meant adventure.

In 1964, when I was about four years old, Dad was appointed vicar of Geraldine, replacing Bob Lowe, who had only recently lost one of his three young children to cancer. The Reverend Lowe was a local identity, and as Canon Bob Lowe, he soon became very well known throughout the nation as a writer, broadcaster and raconteur. Back then (actually, little has changed), Geraldine was a small country town nestled under the fertile green foothills of the Southern Alps. It was a typical rural service town, and the main street was filled with farm supply stores such as Pyne Gould Guinness, Wright Stephenson's, Dalgety's and the National Mortgage Company. On any given day the street would be lined with muddy Land Rovers and the odd Bedford truck, as those who farmed the surrounding district stocked up on supplies. For a small boy with a Land Rover fetish, it was heaven.

Although I was still living in a vicarage, Geraldine was very different to where we'd been living in urban Christchurch. I got in the habit of button-holing the farmers who rolled up to church on a Sunday morning, and asking if I could go home with them for the afternoon and help to feed out or open gates for them on their lambing beat. Long-suffering Christian gents such as Tony Roberts — an easy-going, hardworking salt-of-the-earth Kiwi bloke if ever there was one — would nod, and after the service I would climb into (or onto) their car and be driven back to the farm. In Tony's case, the farm was in the aptly-named Beautiful Valley, one of the last lush tracts of lowland grassland before the road climbs up into the high country. After the typical Canterbury farmer's Sunday lunch — roast lamb and veges — Tony would take me out on his Fordson Dexter tractor. It wasn't long before he began giving me lessons in driving the thing. He had the patience of a saint, which came in handy as he explained over and over the intricacies of letting the clutch out smoothly.

'Ease it more gently, laddie,' he would murmur. 'There's no hurry.'

As I lined the 8-foot-wide tractor up on a 10-foot-wide gateway, his voice would betray nothing of any nervousness he may have felt.

'Whoa up a bit, lad, and keep your eye on the gap,' he would say quietly from the transport tray, and carry on feeding off bales of hay, or snaffling up wet, shivering mis-mothered lambs with his crook to take home and revive in front of the Aga range. At dusk he would dutifully drive me 15 miles back down the valley in his Hillman Husky station wagon to deliver me to the Geraldine vicarage, often enough with a leg of lamb or a dozen eggs as a bonus.

Or I might get a ride with dear old John Bolderstone, a larger-than-life man of the land, who would take me back to his farm with him after a church working-bee on a Saturday and integrate me into his family for the weekend. John had a heap of interesting machinery on his hill-country farm — a Fordson Super Major tractor, a D2 Caterpillar bulldozer, a short-wheelbase Land Rover and a Morris Commercial truck. Lucky!

I remember one occasion when I was out on the lambing beat with John and his sons, Geoff and Mark. As we approached a gate leading into a steep paddock, John stopped the Land Rover, pulled on the handbrake and got out.

'Geoff and I are going to go for a bit of walk around the hill to move a mob. Mark, do you reckon you can take the Landy back home?'

His younger son's eyes shone, but John was looking at me, at seven by far the older and more responsible child. His Mark was about four.

'Ooh, yes!' I said without hesitation. 'I can manage!'

'If you get in trouble, just turn her off and walk,' he said.

As John and Geoff plodded off into the paddock, I settled

myself into the driver's seat, with little Mark offering advice as co-driver. It was a stretch reaching the pedals, but perched on the edge of the seat, I could just manage it. I gripped the Bakelite wheel, still slightly warm from John's calloused hands, selected second gear — even then I knew the golden rule of Land Rovers: if you can't get there in second gear, you won't get there at all — and eased the handbrake off as I slowly let out the clutch. With a lurch, we were in motion. John didn't even look back to check how I was getting on. I crept along in second gear, surging forward in little bursts.

After negotiating a few steep-ish hills, a creek crossing and the odd bog, and without hitting a single gatepost, we arrived back at the homestead, where I took particular care to park the Land Rover neatly in the house paddock.

'Where's Dad?' Nan Bolderstone asked Mark when we came inside.

Mark explained, his mother looking at first astonished, then outraged. When John and Geoff finally got back, she gave him a piece of her mind.

'What do you think you were doing, leaving the vicar's son to drive himself home on his own? He could have broken his neck!' she said.

'He was fine,' John shrugged. 'If he'd got into difficulties, he knew how to turn her off and walk home. Didn't you, lad?'

I beamed back at him.

Another long-suffering, Land Rover-owning saint was Austen Deans, the well-respected Canterbury landscape artist. Often on a Friday night, Austen's older sons Nick and Willie came into town to enjoy the bright lights of Geraldine — the church youth group — and if I spied the Deans's 1956 ragtop Series 1 Land Rover parked outside, I would rush to ask Mum to ring her great friend Liz Deans and ask if I could come and be the eighth Deans boy for the weekend. More often than not the answer was yes, so I would run down the road in the dark to the Land Rover and tape a note onto the steering wheel instructing the Deans boys not to leave town without me. I spent many a late Friday night perched in my PJs and dressing-gown on the cold aluminium rear deck of the Landy as we zig-zagged our way up the Rangitata Valley to the Deans's small farm in the Peel Forest. The trip up had much more to offer than the Geraldine Youth Club: I got used to stopping at various isolated homesteads en route, where 'country parties' — groups of teenage boys gathered to drink beer — were in full swing. I had my first taste of beer at one such event. It was sour to my childish taste and not to my liking, and I think I might have spat it out.

At the farm, I learned a heap tagging along at the heels of the much more capable Deans boys. We would go on fishing, ice-skating or shooting expeditions, or — even better, so far as I was concerned — a 'rough ride' in the Land Rover, where 16-year-old Willie would drive at high speed over the roughest bits of the paddock or up impossible-looking hills, showing off his fearless off-road driving skills while young Michael Deans and I whooped with delight.

One of the unique things about staying at the Deans's place was that all seven boys, plus the odd extra like me, would sleep out year-round on the upstairs veranda, where the beds were all lined up like an open-air dormitory. This often meant waking up in the morning with frost or even a dusting of snow on the end of our beds. It was a bit chilly at times, but you had a grandstand view if a thunderstorm rolled over the plains and when the first light touched the snow-capped summit of little Mount Peel (there are two Mount Peels) just across the way. I learned to shoot a .22 rifle from that veranda. Our target was usually a hapless magpie, but we had to take great care not to shoot towards the skaters on the ice-skating rink at the foot of a bush-clad hill about 200 metres away.

Sometimes the Landy was seconded by its rightful owner, Austen, when he wanted to scout a piece of landscape for another painting of iconic high-country scenes. We would wind up the Rangitata Gorge to the base of the Southern Alps, along the way passing through such legendary country as Mesopotamia, the sheep station owned and farmed at one stage by Samuel Butler, author of the classic novel *Erewhon*. Or we would ford the Orari River in its steep-sided gorge below Mount Peel, and Michael and I would amuse ourselves while Austen painted. We would explore and catch cockabullies — little native trout — and climb hills and roll rocks down the steep slopes. One day, a very cross, ruddy-faced man lurched over the ridge and tore strips off us for rolling rocks onto his eight-wire fence and destroying it. He wasn't such a bad bloke, really: he didn't make us repair the fence (which I would have done in his place), and he even showed us a spot where we could roll rocks without demolishing anything. Once, when returning to this hillside a few weeks after a happy afternoon rolling rocks down it, we found it neatly turned over. The canny farmer had employed our youthful enthusiasm in clearing the stones so that he could then disc it (that is, use an appliance with big cutting discs that turn over the soil and prepare it to be sown with pasture grass) with his D2 crawler tractor.

I learned a lot from Austen about living and camping in the bush when I tagged along on Deans family adventures. On one occasion, I was invited on a pony-club trek with the Deans and Grigg families up into the Lake Heron Station high country, many miles up a rocky valley that connected the Rangitata and Rakaia river headwaters. I was loaned a wise old pony called Waxeye by the Aclands at Mount Peel.

I hadn't done much riding. Waxeye had a pony pad (not quite a saddle, more a woolly sheepskin with a leather strap acting as a surcingle to bind it to the pony, and little cup stirrups to help with rising to the trot), and although I could hardly rise to a trot at the beginning of the four-day ride, by the end I had learned to grip the front of the pad and hang on for dear life when she cantered. At eight, I was by far the youngest in the group of about 20 very grown-up teenagers, and I was keen to help and gain favour whenever possible. One evening, while waiting for the dinner to be cooked on the open fire, I saw one of the older girls, Jenny Greene, carrying a billy.

'Can I help?' I asked. 'What are you going to do with the water?'

'Put it on the fire,' she replied.

I took it from her, and that's what I did.

'Oh, hell, Mark! What are you doing?'

Wreathed in steam and protests, Austen gently explained to me that I was supposed to put the water on the fire, billy and all.

'Never mind,' he said. 'We can just re-light the fire.'

I went away tearfully and hid my shame and embarrassment.

Mother, who often took me on these excursions up into the high country, gave me a very long rein. She didn't believe in wrapping children in cotton wool, and was quite prepared to let me learn things the hard way. One day, she and Liz Deans were shading their eyes and watching me and the Deans kids sliding down a scree slope high above the rocky floor of the Jolie Valley.

'Don't you feel bad about that?' Mum said.

'I feel seven times as bad as you do,' Liz replied. She had seven kids up there with me.

The Deans family were keen skiers, and it was with Austen that I had some great adventures in the mountains, mostly up at Round Hill skifield beside Lake Tekapo. Half the fun was getting there, battling deeply drifted snow on the roads in his Land Rover. On one occasion, we went up to the Ball Glacier, staying in Ball Hut on the edge of the awesome Tasman Glacier. The hut has since gone: with the retreat of the glacier, the hut was threatened by the erosion of the moraine wall and it was removed in 2009. I got my first real taste of mountaineering, carrying our skis around to the foot of the rope tow that used to be situated on the Ball Glacier, a tributary to the mighty Tasman. Along the way I turned my attention to the near-vertical ice fall that is the Caroline face of Mount Cook. I got about 8 feet up before peeling off and landing in a crumpled heap at the foot. Geoff Bolderston came to my rescue, and I needed little of his gentle persuasion to abandon my attempt to be the youngest to climb Mount Cook by the eastern Caroline face.

From Geraldine, it was about an hour and a half by winding gravel road to Burkes Pass, on the far side of which the hills fell away to reveal the majestic sweep of alpine tussock land that was the Mackenzie Country. We often went through to stay with Donald Burnett and his sister Catriona St Barbe Baker on Mount Cook Station across the Tasman River from The Hermitage, which was wonderful for me, because not only did Mr Burnett farm some of the grandest landscape on Earth, but he had not one but two Land Rovers, an ex-army Jeep and a 4WD GMC (General Motors Corporation) tip-truck. I would pester him to be taken along on any of his trips out on the station, such as up to the intake for their private hydroelectric scheme, or distributing salt licks to the stock high on the Burnett Range overlooking the Mount Cook Hermitage airfield.

One day, when I was nine, Mr Burnett gestured to me that I should take his place behind the wheel of the Jeep and drive. As I steered across the river flats, he crouched in the footwell on the passenger's side so that it would appear to my mother, who was trotting along on horseback beside us, that I was in sole charge of the vehicle. She did a double-take as she saw me drive by, grinning, as though I was in charge.

Oh, how I longed to be!

20

CHAPTER 2

A COUNTRY BOY AT HEART

When I was eight, Dad accepted the challenge of being the vicar of Hornby, a run-down industrial suburb of Christchurch. I was sent to board at Waihi, an Anglican preparatory school, occupying an old wooden homestead just a few miles down the Waihi River from Geraldine. There were only about 78 boys in the whole school back then, and, while there were the inevitable dust-ups and antagonism here and there, I made some great friendships in the five years I was there.

I had become accustomed to the wide-open countryside, and although life at Waihi was pretty spartan and tough at times, it was a far better proposition for me than going to a city school in Hornby. The spirit of self-reliance reigned supreme, both a product of and an explanation for the fact that amongst the old boys were people such as entrepreneur and jetboat inventor Bill

Hamilton and Victoria Cross-winning war hero Charles Upham. We were allowed to run pretty wild at Waihi, and positively encouraged to push Nature's boundaries and find out for ourselves the natural consequences. That's the kind of approach that produced men of the calibre of Charlie Upham. Incidentally, I met Charlie once, not long before he died. He was a resident of the rest home that occupied the premises and grounds of Bishopscourt in Christchurch, which had been the seat of the Bishops of Christchurch — my Grandfather Alwyn included — until the upkeep got beyond the Anglican Church. I was visiting a friend of my mother's, Priscilla Neal, who was also a resident, and when I wondered aloud about the possibility of meeting Charlie, they jacked it up for me.

He was pretty frail, but mentally acute. He was lying back on his pillows in bed making conversation with me for most of the time, but when I mentioned my cellphone, he said he hoped it wasn't of German manufacture. A surge of energy went through him, and he sat up rigid in bed. I won't record his comments here, but suffice to say he left me in no doubt at all of his opinion of his wartime enemy!

Most of our extracurricular energies and ingenuity at Waihi were devoted to finding food in some way, shape or form, as we seemed to be perpetually hungry. We were always going down to the river to catch fish in our spare time, or raiding orchards for apples. We also spent quite a lot of time in the old man pines along the boundary of the playing fields, building and occupying tree houses. Your status in the school pecking order was indexed to the size and ambition of your tree house. Soon after I started at Waihi, I attended a 'meeting' of established tree-house owners to work out whether I would be allowed to start building one of my own.

'Why should we let a new bug in?' one of the boys asked, 'new bug' being the dismissive term for a new boy.

David Innes, the son of the owners of Haldon Station where I had been a regular visitor (and, incidentally, star of a then-famous book named *David, Boy of the High Country*), piped up.

'He's got nails,' he said.

The committee exchanged swift glances, and when I produced a handful of four-inch nails, they crowded in greedily.

'Can you get more?' someone asked.

I nodded. Mum and Dad were living only 15 kilometres away, and there was a generous supply of nails in the vicarage maintenance shed. I'd be able to pilfer them when I went home on Sunday leave and bring them back in my sponge-bag.

'Right. You're in.'

Some of the things that we got up to at Waihi would give the various PC brigades in charge of things today conniptions. The school tolerated our arboreal activities, until one day a boy was spied near the very top of a very tall pine, peering into a magpie nest.

'What the hell are you doing up there, boy?' George Everiss, the headmaster, thundered.

'Teaching the magpies to fly, sir' was the retort.

Someone smuggled some old .22 ammo onto the grounds and hid it in the fishing shed, so he could go rabbit shooting at a friend's on a leave Sunday. We got a bit inquisitive about what made a bullet go bang. Someone had heard that if you put a .22 bullet in a vice and wacked the flat end with an old file, it made a good noise. So it proved! Of course, our curiosity didn't end there, and someone brought along a 12-gauge shotgun cartridge as our next subject. For safety, one of the wiser boys (he now has a very successful Angus stud, further proof of his ability to think things through) suggested we should empty out the lead shot first. Then, holding the end of a file on the detonator and bashing the other end with a rock, we set the thing off. The bang was tremendous. We all scattered rapidly in different directions so that the investigating on-duty master, Yogi Brown, would catch some of us, but not all of us. And needless to say, those who were caught proved to have a total lapse of memory when it came to naming the ringleaders.

Another time, a boy decided to do a Tarzan-style swing across the modelling-room shed on the electrical flex from which dangled a single, bare lightbulb. The result, quite predictably, was that the fitting let go and left the wire bare. To save my mate the penalty for this kind of wilful destruction — probably six strokes on a bare bum with Mr Lornie's size 11 sandshoe — I offered to repair the damage. We 'borrowed' a set of insulated fencing pliers from the handyman's tool shed. (It was locked, of course, but we had grown accustomed to scraping away the shingle under the door so we could wriggle beneath it to help ourselves to tools. We always put the shingle back, so we were fairly sure our unauthorised use of school tools went undetected.) Then, taking care that the switch was off, I climbed up onto a chair and went to strip what I supposed was an inert wire.

There was a flash and a loud bang. I was flung across the room, and the school was plunged into the semi-darkness of a dull winter's afternoon as a fuse on the main switchboard tripped. I was shaken, but wiser in the ways of loop circuits and the inadequacy of the plastic insulation on fencing pliers. I had to lie pretty low for a while.

We were free spirits and farm boys, and were given as much freedom to experiment as was compatible with keeping us out of real danger. The approach seemed to work. If I had to name my classmates who had gone on to great things, I wouldn't know where to begin.

I can't say I was an academic success at Waihi. Becoming a woodwork mentor was the highest official accolade I received, but unofficially I gained a bit of a reputation for being handy. One day when the whole standard four class — all 12 of us — was playing up, we were sent outside to pick up a trailer-load of rocks as a punishment. The school was establishing a new hockey field, and, as the piece of land chosen for the purpose had previously been a river bed, there was no shortage of rocks. To expedite the hockey field, the teachers were looking for any excuse to send us outside rock-picking.

Reckoning that we would make shorter work of it if we didn't have to wait for the old handyman/gardener — who was actually known as 'Handy' — to come and shift the tractor and trailer 25-odd metres down the row each time we finished a section of the paddock, I offered to drive the tractor. I was about 10 at the time, and Handy looked dubious. But I pointed out each of the controls and told him I knew how to work them, and he gazed longingly at the shade of a nearby willow, where he spent his time lounging and smoking roll-your-owns when he wasn't driving the tractor.

After a pause, he nodded. So I boarded the tractor, fired her up, selected first gear and let the clutch out. My classmates were inspired by the new efficiency, and set to scampering about rock-collecting with a will as I idled the tractor down the length of the field. The job was completed in record time, and we were all able to race off to play Bullrush on the slightly smoother rugby field. It was my one and only moment of fame and favour at Waihi, and I basked in it.

Like most boys' schools — especially Anglican schools for boys in Canterbury — Waihi was big on sport. I didn't shine much in that arena, either. My cricketing record was short-lived: the first time I walked to the crease, I was bowled out for a golden duck. I didn't get to bowl. I was selected for the First XV in rugby, largely due to the fact that there were so few eligible candidates (that is, old enough and big enough), but my career in this code was pretty short-lived, too. In my first match I got the ball on the fringes of a ruck, and instead of running straight ahead into the pile of bodies — the traditional route, and one which I judged to be futile — I ran around it. Never mind that I made a few yards before I was tackled: I had shown scant disregard for the orthodoxies, as I was told in no uncertain terms (and in words of fewer syllables) by Bruce, the halfback, and his brother Robbie the captain. I found myself relegated to the Second XV. Perhaps

Bruce and Robbie were right: the modern game seems to rate the ability to charge straight into the heavy traffic quite highly, and the Deans brothers became quite famous for possessing it.

I might have thought that I had never achieved anything in my school sporting endeavours had I not recently come across a little mug made of tarnished silver that was inscribed with my name and commemorated a podium finish in the school handicap race — a consolation prize, perhaps. And I was briefly the talk of my classmates' parents after the annual cross-country race one year — a gruelling ordeal by contrast with the safe run-arounds in cultured paddocks they have these days, consisting of a mad scramble across freshly ploughed farm paddocks, through the stony braids of the Waihi River and over open irrigation ditches, all under the watchful gaze of teachers manning three checkpoints. As I floundered across a sea of furrows of rich, dark earth, I spied a bit of snig chain, probably broken off from a set of heavy harrows. Bet someone would pay a bob or two to have that back, I thought. I might even be able to buy a pie or two! So I hauled it from the mud and, hiking up my footy jersey, I wound the heavy chain around my midriff and re-joined the race. If I thought I could sneak it across the finish line, I was mistaken. While others staggered in red-faced and clutching their sides from the stitch, I lumbered across the line, manfully trying to keep the heavy chain wrapped around my hollow tummy in position. Those watching were as impressed as they were puzzled. My parents soon learned that, while I had finished well down the pack, there was no shame in it: I had been carrying quite a few extra pounds of chain at the time.

It was instilled in me from a very early age that if I wanted something, I had to work for it myself. Trouble was, with Dad being a vicar, there weren't as many chances to do wage-earning work for him as others had with their dads. There's not as much room for enlisting part-timers in the cure of souls as there is for shed-hands, or workshop sweepers or the like. Mum used to give me a penny a week pocket money, but it was conditional on keeping my school uniform neat and clean. With my lifestyle this wasn't easy, and therefore this couldn't be regarded as reliable cashflow.

One of the best earners for kids in the days of refillable glass packaging was collecting empty drink bottles and returning them to the dairy to collect the deposit. A small fizzy-drink bottle returned to the dairy fetched three pence, a large one sixpence, and a beer bottle tuppence. The holy grail was a half-G, the clear, half-gallon flagon that men (it was always men) used to carry home, both men and bottles filled with beer, after the pubs closed at six in the days of the 'six-o'clock swill'. These were worth half a crown (two and a half shillings). This might seem like small change, but you have to bear in mind the contemporary purchasing power. An ice block on a stick was tuppence, and a really good double-header ice cream in a cone was sixpence. My first bike cost the equivalent of 30 half-crowns, second-hand.

It didn't take me long to work out that at precisely 6.05 every evening (except Sundays), there would be a mass exodus of drinkers loaded down with crates of 750-millilitre beer bottles of

Bavarian Bitter or DB Green or flagons of draught tucked under their arms, intent on carrying on. Most family men went home to drink and spend time with their wife and kids, but the young and rebellious would zoom off in their Mark 1 Zephyrs to a nearby park or reserve and party on into the night. They weren't exactly tidy Kiwis. I can't have been much older than six when I realised that if I got up early enough on a Saturday morning (about 4.30am), I could get to the scene before anyone else and clean up in every sense of the words. I didn't own a watch, but with the first hint of pink in the sky, I would dress quickly (that is, pull my very second-hand, home-spun woolly jersey over my pyjamas and poke my bare feet into gumboots), take my sister Celia's pram (she was two) and set off on my rounds of the various parks and party spots. The younger rebels, who hadn't reached the drinking age of 20, used to congregate around Robbie's Milk Bar, 200 metres up Talbot Street from my house, and there was nearly always a good haul of fizzy-drink bottles in the gutters or tossed into darkened corners. By six, as the other solid citizens of Geraldine were still snoring in their beds, I would proudly present my pram-full of bottles to a local dairy for my reward. I could never understand why the shopkeeper wasn't pleased to see me at that time and always 'firmly encouraged' me to come back much later.

One morning, in the dawn light, I saw another boy, quite a bit older than me, pushing a cart along on the opposite side of Talbot Street. I had competition! His business model was the same as my own, and for a few Saturday mornings we raced each other to be first to the best spots. Soon we tired of this destructive cat-and-mouse competition. We introduced ourselves and agreed to a partnership, divvying up the spoils. He was older than me and much worldlier. He even bragged that he had bought himself a tractor with his earnings.

One day, he told me to tag along and led the way to the motor camp. He selected a cabin whose occupants were probably off somewhere racing their bodgie car (a hotted-up Mark 1 Zephyr or the like). Holding a finger to his lips, he produced a screwdriver and unscrewed the lock from the door. Inside, there were two or three crates of beer. My partner in what, young as I was, I knew to be crime, took a bottle and opened it. He took a swig, made a face and passed it to me. I didn't like it, either. So we emptied the bottles outside and added them to our collection. I was feeling queasy, and it wasn't from the beer. The feeling didn't go away after we had cashed in the bottles at the pub. I was grateful a few months later when Mum announced that she was off to Britain to visit her relatives and was going to take me along. I was able to extricate myself from the petty-thieving racket, and I didn't renew the contact when we got back from the trip.

Still, all-in-all, the bottle lark was a valuable life lesson. It turns out that the early bird really does get the worm (even if I have subsequently learned that it's often the second mouse that gets the cheese).

<hr />

After Waihi had done its preparatory work on me, I progressed to Christ's College (which we always called College), attending as a day boy while living with my parents in the Hornby vicarage. City life didn't appeal much at all, and I longed to be back in the country. Happily, Gordon Harper, one of my great friends from Waihi, lived on a cropping farm at Carew just outside Geraldine, so I could get out of town now and then, especially in the summer holidays. To my delight, I found that Gordon's parents, Eric and Ann, had a very high opinion of the capabilities of 12-year-old boys. The Harper family had a go-kart powered by a 4-stroke 6-hp Honda motor, and we used to slide this around on the muddy paddocks. I thought that was cool enough. But then, when it broke, Eric patiently taught me how to use an arc welder to repair it. And while other farmers had been happy enough to let me drive a tractor, Eric actually put me to work, using the tractor to do grubbing and heavy rolling. This was heaven!

By the end of the first summer I spent out on their farm, I had done a fair bit of work with the Harpers' 165 Ferguson. One afternoon, Mr Harper pointed me to a patch that needed grubbing and to the much bigger and newer Massey Ferguson 188. Who knows what gave me the bigger buzz, being in control of all that power, or being trusted with it?

By the time the next summer holiday rolled around, the Harpers had a brand-new 80-hp Massey Ferguson 1080. I secretly hoped Mr Harper would let me drive it, and he did. I was careful at first, but by the end of a day's grubbing I was handling it without a care in the world. I was just putting it away when there was a bang from over my head. I stopped, killed the engine and climbed off. I had clipped a low beam in the shed, and the exhaust pipe was sitting at a slightly rakish angle. I thought that would be the end of my adventures with the Harpers' agricultural machinery but, apart from a bit of muttering, Eric Harper wasn't fazed. A few blows with a sledgehammer restored the exhaust pipe to its former, upright position, and everyone chalked it up to experience. I'm sure I became a better driver because of it.

I hadn't long turned 14 when Eric Harper asked whether I'd like to join the harvest team, taking turns with Gordon driving the 12-ton Morris tip-truck alongside the header collecting the wheat on the run. After each run and when the truck was full, we would then drive 3 miles to deliver the load to the silo.

The truck was a beast of a thing to drive. In the absence of effective power-steering, you had to bodily fight the huge wheels around, using both hands to haul down on one side of the steering wheel. You were perched right behind the hardworking engine, and, because harvest time was high summer and the blistering nor-wester seemed to be blowing non-stop, it was exceedingly hot. But Gordon and I thought we were pretty cool, priding ourselves on smoothly double-declutching the crash box on the downshift without grating the gears (much) and slickly working the 2-speed Eaton axle. In our minds, we were behind the wheel of a Kenworth. If we'd had an air horn to toot, I'm sure we would have worn it out!

On one trip, when it was Gordon's turn in the driver's seat on the way to deliver the load of fresh wheat to the silo, Eric waved me up onto the Massey 650 header to sit beside him on the open driver's platform so that I could learn that skill, too. After we'd done a few rounds, creeping along the rows as the tall plants disappeared beneath the table containing the roaring, seething machinery in front of us, Eric stopped.

'Take over for a bit, would you, Mark? I've got to nip back to the yard to check something.'

Awestruck and thrilled all at once, I slid along the seat, grasped the steering wheel and worked the actuating levers and controls. I took it slowly and cautiously at first, conscious of the heavy responsibility with which I had been entrusted. But I soon loosened up, and while I was doing a spin turn at one of the tight corners at the end of a row, there came a shuddering thump and a cloud of dust from the front of the machine. A humped-up border dyke was hidden beneath the standing wheat. Having failed to notice it, I had dropped the front table right on top of it, and, instead of the stalks of wheat, the flails had scooped up a load of dirt and stones. Mortified, I threw the thrashing table out of gear and reduced the engine down to an idle to inspect the damage. There was a load of dirt and stones amongst the flails, and, without stopping to think, I climbed in to remove them to stop the grain being contaminated. It was only later that it occurred to me that if Eric (or anyone else for that matter) had come along and found the header idling with a crop to harvest and had got on with the job, I would have been mincemeat. That was one of my lives gone, right there! With no obvious damage to the header, there didn't seem to be any point in letting on about the incident to anyone. That particular episode has been embargoed for 42 years.

I am full of admiration for the people who, like Eric Harper and the rest, held young people in high enough regard to trust us with jobs like this. We rewarded his trust with a fierce determination to not let him down. We learned lots of skills at an age where learning comes easy and you're young enough to give the machinery the full respect it deserves. It frustrates the hell out of me to hear young people say they can't do even the simplest jobs, when Gordon and I were driving trucks and tractors and operating harvesters at the same age.

※※※※※※※

Perhaps due to my as-yet-undiagnosed dyslexia, my years at College were nothing to write home about, at least not from an academic perspective, and again I wasn't exactly the jewel in the school's sporting crown, either. But having spent my childhood obsessed with machinery, I was discovering that I had a considerable mechanical aptitude. As we got older, my classmates would occasionally be given a clapped-out car or motorbike by their well-intentioned parents, and, with Saturday night's social activities looming and lacking the basic know-how themselves to get or keep their vehicle running, my mates would come to me. Quite often it was nothing serious — nothing that filing a set of corroded ignition points or back-flushing a blocked fuel filter couldn't fix. Then a boy who had summoned me with pleading eyes would turn the engine over at my invitation and I would be rewarded with an expression of unfathomable gratitude as it caught and ran, sweet as anything. It helped, too, that my charges were modest, especially when compared with the quotes supplied by opportunistic garage-owners only too happy to help College boys with their perceived affluence. Thus it wasn't only because I was from a noted clerical family that I became generally known as 'Rev'.

By now I had transport of my own, which possibly enhanced my reputation. My first set of road-legal wheels was a 50cc Vespa scooter. It was 8 miles from home to College, a fair distance to bike on all but the best days, but there was no point in complaining to Dad about that. However, when I followed Dad into his sport of rowing, the need to get to the college's rowing sheds at Kerr's Reach added another 5 miles to the trip, and even Dad accepted that this was a bit much for a growing lad. So he weakened, and allowed me to purchase my very second-hand scooter for $80. I already had a Suzuki TM125 motocross bike; but without all the bells and whistles — lights, indicators and so on — this couldn't be ridden legally on the road. The Vespa hardly fitted the image that a keen motocross rider sought to create, but I soon found that the rear-mounted engine made for a weight imbalance that enabled it to be 'wheelied' — ridden with the front wheel off the ground — for quite a distance, often the length of a rugby field, to show off!

Meanwhile, and much to Dad's disappointment, I didn't get far in rowing. However, the headmaster, Tony Brough, soon learned that I had a bit of a knack for getting the best out of an engine, and so he appointed me to be CBD (coach-boat driver), in charge of the powerboat that trailed the crew on the water. Who knows whether it was any consolation to Dad when Tony Brough remarked to him at the College Leaver's Function that the powerboat would never go so well again now that

I was moving on. And while College didn't award a prize for mechanical aptitude, the Latin master, Jimmy McBride, offered a vote of confidence when he asked me to re-tune his lethargic Volkswagen Beetle. My efforts with his Beetle paid off more obviously than his efforts with my Latin. It only occurred to me later that if I had reported to him that the car was partly suffering from *tardius ignitionem* and simply fixed by advancing the retarded ignition timing, my practical demonstration of the importance of Latin might have paid an academic dividend. No such luck.

My first legal four-wheel road vehicle was a 1956 Vanguard truck, which my mother and I purchased for $100. (I negotiated the deal, she paid.) I was feeling the need to cart my motocross bike to race meetings, and I had also identified certain synergies with the need that my horse-crazed mother and sister Celia had to drag a horse-float to pony-club events. I could have paid for the truck out of my holiday savings, but the joint venture made far more sense. 'Old Truckie', as he was known, had many adventures. He became a common sight around Christchurch on Saturday nights, where he served as a party wagon, chugging about in search of action with anything up to eight lads and ladesses leering it up on the tray. In winter, Truckie became an essential part of the skiing legends we created, plying up and down narrow, snowy skifield access roads with the crew on the deck getting a head start on the *après* festivities. The downside of Truckie was that he was a prolific drinker, managing only 17 miles per gallon on flat, open roads with a tail-wind, and so an undesirably large amount of the money that we would have preferred to spend on DB Green was turned into petrol and tipped into his tank. He was really cramping our style — until an electrically-minded friend helped to rig up a hidden switch in the fuel-gauge circuit. Whenever Truckie was approaching a fuel station, the switch could be flicked and the needle would drop to empty, enabling us to make a polite, but definite, application to the party committee on the back for funds. With doubt suddenly cast on our keenly anticipated arrival at the next party destination, we invariably found those on the back willing to cough up, and that usually helped cover our own drinking expenses as well.

On one occasion, Truckie's drinking habits got in the way. I was returning from a week's skiing at Tekapo with one of the locally notorious Baxter twins (Paddy, I think, but it might have been Mark). The budget was as tight as ever, but while we had calculated the fuel item to the nth degree — we had $8 to get us 140 miles — we hadn't reckoned on Truckie's lately much-enhanced appetite for engine oil. By the time we had reached the Kakahu saddle about 10 miles west of Geraldine, the needle of the oil pressure gauge had slumped to zero and the engine had acquired a slight rattle. A change of plan was essential. I hitchhiked the rest of the way into Geraldine, and used our lunch money to purchase a gallon of engine oil. That somewhat revived the pressure gauge, and got us as far as the garage in Geraldine, where we topped up with oil again. But by this time, the slight rattle had developed into a more pronounced knocking.

The garage man listened to it in alarm.

'Where are you boys heading?' he asked.

'Christchurch,' I replied.

'You're dreaming!' he laughed.

Belching smoke and fumes, we pressed on grimly. On the main straight on Arundel Road between Geraldine and Ashburton, there was a final, loud bang. I put the clutch in, and Truckie rumbled and squeaked to a halt. There was deathly silence, apart from the pinging of hot metal. We climbed out and inspected the damage. Truckie had refunded our lunch money in the form of a big, hot, black, oily puddle on the hard shoulder, and retracing the trail to the scene of the fatal event, we found half a conrod buried three inches into the tarseal.

But loyal to the last, Truckie had chosen a convenient place to suffer a severe internal haemorrhage, just a few miles from the Harpers, from whom we borrowed their family Holden station wagon. Truckie was dragged into his economic afterlife at the end of a 12-foot length of old, pink ski-tow rope. For $30 I picked up a water-logged and seized but salvageable motor from the local car-wrecker and, with the help of a few of the lads, Truckie was back on the road in about the same time as it took me to write an English essay (about two weeks). A few months later, Truckie was sold to a fencing contractor in North Canterbury for $200.

Not that my life was all play and no work. During my time at College, I was lucky to have some well-paid holiday jobs. The May holidays were always spent helping Donald Burnett with the autumn muster at Mount Cook Station, where I had learned the skills of mustering the Merino sheep from the high mountaintops from an early age. You don't always need a dog with the wily, flighty Merino, so it's essential that you quietly position yourself in such a way that you can avoid the need to run uphill if a mob breaks back. With 32,000 acres to muster, Mount Cook was about as good a proving-ground as you could ask for, and the lessons I learned there set me up for a life in stock management. Donald was a very conservative farmer — he favoured low-input, low-intensity methods, and practically everything he was doing on the farm had been done that way for the better part of a century — but he was a great teacher with it.

At the end of the fifth form, I got a summer holiday job at Steele's Motor Assembly in Hornby. Steele's assembled Toyota Land Cruisers, and my job involved carefully sorting the Firestone SAT (Super All-Traction) directional mud tyres so that the treads went onto the vehicle in the right direction, as well as unloading containers of springs from Japan and making sure the various short- or long-wheelbase models were supplied with the correctly weighted springs. It was great experience, if a bit tedious. The money was very good: I was embarrassed to find I was earning more per hour sorting tyres than my father got for helping the Lord sort the spiritual wheat from the chaff. Undoubtedly one of the best pieces of advice I gained from my time there was when the manager, John Pearson, pulled me aside one day and asked how I was enjoying the job.

'I love it,' I replied.

'No, really. Do you enjoy it? Be truthful, now.'

I admitted it was a bit tedious and boring at times.

'Right,' he said. 'So go back to school next year, work hard and get UE or you will be doing this for the rest of your life!'

I was still riding the Vespa to work — a lean, 6-foot lad proudly wearing a fully decorated motocross racing helmet, but hunched over the handlebars of an underpowered scooter designed to be ridden by petite Italian ladies in stiletto heels. Some of those with whom I was working were my age, but they had cool, flash cars, such as hotted-up Mark 4 Zephyrs. There was a certain allure to chucking school in, taking the money and spending it on toys that I could use after-hours.

But one day, walking to catch a big red bus into Cathedral Square, the incumbent Bishop of Christchurch, Alan Pyatt, stopped his bike to have a yarn.

'What are you up to, nowadays?' he asked.

I made a face. 'Dad says I have to stay at school until I get UE' was my adolescent answer.

'Wise man, your dad,' Alan nodded. 'You'll find in later life that UE is a very easy burden to carry. Good luck!'

He pedalled off to spread the good word elsewhere, but I never forgot those words. Indeed, I've had occasion to quote them to my own sons in more recent years.

CHAPTER 3

A SLIPPERY SLOPE

For me, simply being in the snowy mountains was a drug. So far as I was concerned, a wild blizzard was more exciting than a nice sunny day. Strange as it may seem to many, one of my greatest pleasures is to be driving in fresh, deep snow. It's almost as much fun as skiing in fresh snow — provided, of course, I'm in a suitable vehicle. The profit I made on the sale of Truckie allowed me to upgrade my wheels to a bright orange 1962 VW Beetle. It was soon highly modified to become a 'cool-as-possible' ski wagon, with raised suspension, extra-grippy European snow tyres, go-faster Starsky and Hutch pinstripe and super-powerful halogen driving lights, and was plastered with every ski sticker I could lay my hands on. The 'Starsky Bug' quickly wore out a succession of engines, too; quite possibly due to my conviction that it could go anywhere a 4WD could. I prided myself on the places I could get with my Beetle. I'm sure my friends initially placed bets every time I announced my next ski destination in a deteriorating mountain forecast, but soon there weren't many who would bet against me. The Beetle's reputation for getting up snowy roads when others had given up eventually met its Waterloo when Mark Baxter and I spent most of a day we'd rather have spent skiing retrieving it from over the edge of the Fox's Peak skifield road.

Someone recently asked me how I learned to drive a 4WD off-road in slippery conditions. I replied that if you learn to drive a 2WD in slippery conditions, driving a 4WD seems a whole lot easier.

My love of skiing and the mountains presented me with another fork in the road. Back in the early days, Dad's salary as a vicar didn't run to expensive skiing holidays, so to enable us to spend a week each winter on the slopes, Mum used to act as Camp Mum in the Tekapo Ski Club hut, for which the quid pro quo was free accommodation for her and me. Assuming responsibility for 24 teenagers living in an A-frame hut above the snowline with no electricity wasn't everyone's idea of a holiday, but Mum never complained. Still, when I was about 14, Mum decided I could fend for myself and arranged for me to have a ski week at Tekapo on my own. To get there, I had to hitchhike from Christchurch to Tekapo. In those days, a child's ski pass or tow ticket was 50 cents per day. You could generally count on the weather to take a day or two out of the reckoning, so even though I only had $2 in my pocket, I thought it would be enough to pay to ski for four of the seven days. Trouble was, this particular year we had a perfect week and my money was gone by Friday.

Not content to sit around and watch my mates (who had much more generous means) having all the fun, I walked up the tow line to where one of the skifield staff, Don Hunt, was shovelling snow.

'Can I give you a hand?' I asked.

He nodded to a shovel leaning nearby.

After half an hour of digging together, he paused.

'Why aren't you skiing?' he asked.

'Run out of money,' I replied. 'Can't afford a pass.'

'Well, if you finish that shovelling job, come and see me tomorrow and I'll see what I can do.'

When I rocked up to him the next day, he presented me with a tow pass. For the next few days, that was the pattern. I offered to help and did anything they asked me to do, and in the process I seemed to earn a place on the staff. The following year, I wrote to Karl Burtscher, who ran the field, and asked whether there was anything I could do to help during the holiday that would let me earn my skiing. He wrote back offering me a job in the ski hire, setting up skis and boots in the morning and putting them back in the afternoon when the day's skiing was over. Between shifts, the middle part of the day was my own, and I spent it skiing. It suited me perfectly, and for a 15-year-old to be integrated into the wild group of 20-year-old ski bums who comprised the rest of the staff was both exciting and highly educational!

I didn't confine myself to the ski-hire role, though. I offered to help with anything and everything going, including cleaning the loos and carting the rubbish away. The latter job meant loading the rubbish bins onto a trailer behind an old Series 1 Land Rover and taking the load down the steep, snowy field road to the dump far below. On my first run, I learned soon after I set off that the Landy had no brakes. That was a bit disconcerting! But Austen Deans had taught me well. I used the very lowest gears — the fastest you can go in a Series 1 Land Rover's first-gear low range is walking speed — and on the extra-steep sections I used the water table to scrub off speed, dropping the offside wheels of the tow vehicle and the trailer into the slush of the shallow ditch. This usually slowed me up so much that I actually had to keep a bit of power on to avoid getting stuck. There was a handbrake, of course, but you should never use this for slowing; only for parking.

In this way, I got to the bottom and to the dump safely. Getting back up was a breeze, and I was both surprised and delighted when, on payday later in the week, Karl handed me a cheque for $16, enough to buy two months' worth of tow tickets!

These days, WorkSafe would have kittens if it learned that an immature driver had taken a 2-ton load down an icy skifield road in a vehicle without brakes. But it's amusing to think that I am now paid handsomely to teach police, snow-plough drivers, ambulance personnel, skifield staff and bus drivers to drive on snowy, icy roads. How do they think I learned? A smooth sea never made a good sailor!

I liked this business of getting paid to ski. I realised that I could take it a bit further if I enrolled in the Ski Patrol training programme. It was a long and intense course, and many who started it didn't follow it through. But after a few years of tests and examinations, I graduated and earned the right to wear with pride the bright red jacket with white crosses all over it. I spent my first few days as a rookie ski patroller at Porter Heights, a skifield at the entrance to Arthur's Pass, to the west of Christchurch. I was grateful to be given free skiing and a free lift ticket for a friend — I seem to recall giving these to my sister, Celia — in exchange for skiing around and being a sort of traffic cop on skis. I soon joined the nearby Mount Cheeseman Ski Club, and spent many enjoyable weekends and ski weeks as Duty Ski Patroller.

My great friend Jo Pasley would stock up on groceries and the 'beer essentials' and travel out from Christchurch in her VW Beetle to Kirwee, where I was working for Alister and Liz Frizzell on a cropping farm at the time. We would load up my newest vehicle, 'Jumbuk' — a (very much modified) Series 1 Land Rover — in the dusk and head up the mountain. Many times it was an exciting trip up in the dark, punching our way through snow drifts to reach the club hut at 1500 metres above sea level. Of course, such an exciting trip would require debriefing with the rest of the crew up on the mountain with much Blue Nun, Barker's Negus, glühwein, rum and Coke, or (if you were super-trendy) Bacardi and Coke or Canadian Club and dry. The drifts we smashed through to get to the hut would get taller and taller at roughly the same rate as the stories. On a really good night, there may even have been mention of seeing a Yeti or a moa. (The latter wasn't regarded as completely far-fetched at the time, as Paddy Freany, the publican at the Bealey Hotel, just 15 kilometres further up the valley, had recently reported seeing a moa. Never mind that it just might have been a publicity stunt for his under-patronised hotel.)

The following morning, any claim to have actually seen such an animal would be hastily and vigorously denied.

***

One of the things that may have helped me get the job at Cheeseman was that I was able to drive a bulldozer reasonably well. This meant that Kevin Hobson, the skifield mechanic and road-clearing man, could have the weekend off when I turned up, because I could take over from him on standby in case of big snowfalls.

One Saturday morning, I was shaken awake at 5am by the skifield manager, Steve Feedom, to hear him announce that there had been a big dump of snow overnight, and I would need to be up and onto the bulldozer pronto so that the 7am snow report over Radio Avon could proclaim the field to be open and the road cleared. My head and guts felt in no fit state to do this, but a sickie was off the menu. I refused to acknowledge I was hungover — I was far too young and tough to suffer such an inconvenience!

A cup of coffee and half a cheesecake was all I could manage for breakfast, working on the theory that the cheesecake would line my fragile tummy and, should the worst come to the worst, it would be relatively painless as it re-emerged.

So wrapped up in every bit of warm clothing I could find, I fired up the D4 Caterpillar just as the first slivers of light on the horizon warned of the breaking dawn. It was refreshingly cold. I didn't have a camera, but the pink and orange hue that brightened the leaden sweeps of freshly fallen snow made for a picture that will be in my memory forever. Although the fall had only been about 25 centimetres or so, it had drifted quite a bit, and, by the time I had worked my way down to where the road plunged into a long cutting through a spur, the drifts were higher than the blade on the D4.

With the bulldozer's blade angle set to windrow (that is, to wipe sideways and push the snow out over the edge), it was a fairly simple job to clear a track down to the cutting. But when I reached the pylon of the access tow, I had to turn inwards, away from the edge, to get around it. As the snow was no longer being cleared off the side of the road, this meant collecting an extra big blade-full in front of the machine.

Not a problem, I thought. I'll just turn out sharply once I'm past the obstacle and dump that blade-full over the edge.

What I hadn't reckoned on was that the shaded ground was frozen hard, and as soon as the pushing weight was off the tracks and the tracks were side-on to the hill, the grousers (the sharp steel bars on the tracks that grip) turned into pretty fair imitations of ice skates. The result was that 8 tons of dozer on 24 sets of skates suddenly shot sideways downhill.

I wasn't hungover anymore, although I noted with a curious detachment the astonishing sideways speed that the dozer acquired.

To an observer, it would have seemed as though disaster was imminent. Surely gravity was going to win this one, and nothing could stop the dozer's sidelong plunge off the side of the narrow mountain track, where a 500-foot drop yawned?

But there were no observers, no one to see what happened next.

This could get untidy, I thought, as I quickly selected reverse and applied some power. The inside track churned and, as soon as it made contact with the snow bank at the inside road's edge, the dozer pirouetted around to face downhill again and stopped dead.

Very balletic, I thought.

My heart was pounding, and I had that odd, metallic taste in my mouth that adrenaline gives you. The alternative possibilities that hadn't occurred to me in the heat of the moment now played out before my mind's eye, and I found I was shaking a bit. It took a moment and a few deep breaths before I was ready to proceed with the job in hand. By the time I had cleared away most of the drifts on the cutting, a group of five vehicles had managed to get to the bend at the bottom and the owners were standing, watching the dozer's progress in opening the road for them. The first car in the queue was a very new, fastback, flared-guard Fiat 132 Abarth, one of the seriously cool cars of the time and a top performer in the world rally scene. Its owner was decked out in the latest, brightest ski fashion, and he had a seriously good-looking ski bunny at his side. Some people have all the luck, I thought.

I decided I should make one last sweep of the road before I could indicate to them that I thought it was safe to try to ascend, and having mastered the old sideways-sliding, runaway-dozer trick, I thought I'd treat them to some early-morning entertainment. Pushing the last blade-full of snow over the edge just after the tow pylon, I let the dozer skate sideways a bit longer than last time (about 30 metres), knowing it would track perfectly sideways on its 24 sets of ice skates. Then all I had to do was reverse for a metre or so and the machine would slew around and stop dead. It was all perfectly under control from my perspective. But, as it turned out, the sight of an 8-ton D4 apparently sliding completely out of control straight toward his new mechanical pride and joy was too much for Mr Cool, who panicked, hopped into his Italian masterpiece and promptly backed into the car behind.

Oh dear, I thought. Might have pushed the entertainment factor a bit far on that one.

I concentrated on clearing the last bit of snow bank to enable them to proceed up the hill.

There was no major damage to either car, I noticed, but I suspect Mr Cool had suffered a pretty serious dent to his ego.

I got a chance to atone for this little episode on my way back up the hill when I noticed a Mini sliding across the car park with its back wheels locked up. The rear brake shoes had plainly frozen onto the drums, so when the handbrake was released the wheels remained locked. Because it was front-wheel drive and had chains on the front, the driver hadn't noticed, because so long as he kept in a straight line the rear wheels would follow obediently enough. But I knew what would happen as soon it turned out of the car park. Sure enough, as it reached the steep downhill section, it commenced a series of beautiful, linked 360-degree spins. Sensing that the driver didn't know how to regain control of the wayward Mini, and realising that the situation was turning rapidly from comedy to catastrophe, I managed to scoop up two blades-full of snow and build a bank across the road. The Mini pirouetted neatly into it with a puff of soft powder and disaster was averted. The driver was very grateful for my presence of mind, and vowed never to leave the handbrake on again when he parked in the mountains.

When I left school I decided to do the Diploma of Agriculture course at Lincoln Agricultural College, just outside Christchurch, which required me to gain two years of experience working on a farm. But before I embarked on that, I decided to apply for a job with the professional ski patrol at the Lake Ohau Field in the Mackenzie Country. I was delighted to get the job, and I spent an impatient autumn working for BA Turner Contracting in Christchurch, doing various truck- and tractor-driving jobs, and waiting for the ski season to start. The snow finally came in August, and I was terribly excited as I set off south to spend the winter living with a bunch of international ski bums at the 'Grottoage' (as we called the main accommodation) at Ohau.

Many are the stories from that winter, but two particularly stick in my mind. Among the others working in the ski team was Pete, a trainee dentist with a decidedly practical bent. Pete decided we should build an outdoor sauna beside the Glen Mary

Ski Club. He designed a very well-engineered building, complete with a wood-fired stove packed with greywacke rocks and three-tier seating inside. My memory is hazy on this point, but I seem to recall that the timber for this magnificent structure was 'borrowed' from the Ohau A power project (it was returned at the end of the season!). To complete the 'Swedish' experience, we dammed the snow-fed creek beside the sauna to provide an icy plunge pool. Once we had got the wood-burner lit (Pete used to have to put on his scuba tank to do this, as it was very smoky inside until the heat built up and the flue began to draw), it worked very well, and the female lodge guests whom we invited to strip off and share our sauna doubtless counted themselves very lucky.

In about the second week of my professional ski-patrol career, I faced my first real test. Dropping into the steep gut on the south side of the Ohau T-bar, I saw a couple huddled on the snow. As the light was fairly flat, it took a while before I realised that one of them wasn't lying down in the snow out of choice. Here was my first injury!

I felt an instant flick of adrenaline, but reminded myself to be calm and initiate the 'skier incident process'. I pulled into a tidy but definite parallel stop 5 metres uphill and 5 metres to the side to avoid showering the couple with snow. Kicking off my skis and placing them in a cross 3 metres from them, I trudged down to them in my ski boots.

'Hi,' I said, 'I'm Mark.'

I was confident that my bright red jacket and orange first-aid bum-bag would explain what I was there for.

The patient was a fit-looking middle-aged man.

'I'm guessing that because you're lying in the snow rather than skiing on it, you've got a problem?'

'I think — in fact, I know — I've hurt my leg,' he said calmly.

I was kneeling beside him by now, and I could see his leg was twisted at a very strange angle.

'I'm going to have to try to get that ski boot off,' I said, and he nodded. His wife looked far more anxious than he did.

Trying to release his boot from the binding, I was as gentle as I could be, but he winced visibly. This gave me a sneaky feeling that I was dealing with more than just a twisted ankle.

I explained that, with his wife's help, I was going to try to gently apply traction to his leg and put it into a splint. I didn't tell him that this was the first time I'd dealt with any kind of injury, let alone a broken leg. After all, the importance of being reassuring is one of the first things they teach you.

As I was applying the splint, Brod Wigley, another member of the skifield staff, spied my crossed skis and came to ask what he could do to help.

'The cascade sledge would be handy,' I said casually. Brod set off to bring it from the top T-bar where it was stored. By now, the crossed skis and the sight of a distinctively dressed ski patroller huddled over a body lying in the snow had attracted a few helpers, so we had plenty of manpower to assist in lifting the patient onto the sledge.

'We'll be as gentle as we can,' I told him as I strapped him in. He nodded, pain pinching the corners of his eyes.

We skied him down the rest of the field, over the steep shirt-front slope to the patrol room, where with the help of three others we transferred him to the bed. There, we gave him a bit of pain relief to make him comfortable.

Barry Ems, the skifield manager, let the tyres on the staff Land Cruiser down a bit to soften the ride as he carted the patient down the hill, and I filled out the paperwork that would accompany the patient to hospital for an X-ray. It was all the standard stuff: name, address, age ...

'Occupation?' I asked.

'I'm a doctor,' he said gently.

My pen paused as I digested this information.

He must have sensed my anxiety, because he told me I had done a very good job. 'Well done!' he said.

I tried to keep a neutral, professional expression on my face, but it might have slipped a bit, as relief and gratification fought for control. I don't recall hearing from him again, but I have always taken that to mean that my first-ever first-aid procedure had contributed to a happy outcome.

I loved living and working on the skifields, but I didn't need my parents' constant reminders to tell me that it was a great lifestyle but no kind of career. Contrary to their fears when I put my Lincoln studies on hold, farming remained my primary aim.

CHAPTER 4

WORKING FOR A JOB

The two years' practical farm work I had to do for the Lincoln Diploma of Agriculture was clearly going to be a different proposition to the holiday-type jobs I had done for many years. I placed an ad in the farming pages of *The Press* seeking work, on the theory that I would be in the position of a chooser amongst offers rather than a beggar for a job.

Soon after the ad appeared, a very friendly farmer from Cust, 30-odd miles north-west of Christchurch, rang up and offered me a job on his small (400-acre) mixed cropping farm. I gave it a go, but the place just didn't have enough to offer. The farmer kept giving me menial jobs, and pointedly ignored my insistence that I was far more capable than he was giving me credit for. One day, we were building a shed. The farmer set me to work yanking nails out of old boards while he tried to start a chainsaw to cut the boards to length. After tugging away at the recoil cord for a few minutes to no effect, he set off for home to collect a handsaw. I quickly removed the spark-plug, cleaned the oil and soot off it, replaced it and fired up the saw without too much difficulty. Then, figuring it was running a bit rich, I removed the air filter and gave it a good rinse in petrol. It ran nicely when the filter was dried and replaced. I set to work cutting the boards, expecting to be showered with praise and entrusted with slightly more challenging work from then on, but I was wrong.

'What the hell do you think you're doing?' the farmer growled when he got back.

I started to explain.

'Who told you to go mucking about with my chainsaw?' he huffed. 'You're supposed to be de-nailing that timber!'

After that, he seemed to go out of his way to find simple, boring and repetitive jobs for me.

I told Mum about my troubles, and she reckoned it was a bit like trying to fit an exuberant Labrador puppy into a chihuahua's kennel. The farmer had reached the same conclusion, and one day he folded his arms across his chest and fixed his gaze on the Main Divide to the west.

'Not sure it's working out, Mark,' he said.

I nodded, but didn't say anything.

'It's a little place,' he carried on. 'It's pretty clear there's not room here for both of us.'

He turned and looked me in the eye.

'And I'm buggered if I can see why I should be the one to leave!'

We both laughed, and parted friends (I hope). And in the end, he did me a huge favour, as I managed to get a job at the opposite end of the farming spectrum on Okuku Pass Station, at North Loburn in the Canterbury foothills. The 15,000-acre block stretched from the rolling downland of the Loburn just north of Rangiora up and over the 700-metre Okuku Pass and into the head of the rugged, undeveloped Lees Valley north of the Karetu Range.

Okuku was owned by the Department of Lands and Survey, and was going through a major development phase at the time, with the thick gorse being sprayed out and burnt off, and the land either planted in pine trees or worked up by crawler tractors and giant discs to sow into improved pasture.

My new boss, Linny Morris, was a bit (actually, a lot) of a cowboy, and although he was short, he was strong, and cultivated a tough-guy reputation. He liked to be known as 'the Great Linno', and he had a few testosterone-fuelled stunts calculated to demonstrate his strength and general ruggedness. He would tear telephone directories in half, for example, or take big bites out of drinking glasses. Going to the pub with him was always an edgy affair, because he'd lean there, twitching his biceps and scanning the room for the least sign of trouble — he took great pleasure in single-handedly clearing out unruly bar-rooms. All this, and he was a celebrity, too, on the strength of being a top buck-riding rider at local rodeos and having appeared in a TV advertisement for Moro bars.

The Great Linno took it upon himself to toughen us up, too. Before we were allowed breakfast in the mornings, we had to perform various tests of strength and fortitude, such as lifting a full-sized horse-shoeing anvil above our heads.

A tall, lanky, overconfident and highly opinionated ex-Christ's College boy was not what was required in the job description as a junior shepherd, and for my part it was a rude awakening to discover that formal education counted for very little in the rough-and-tumble environment of a big high-country station, where the power of the bare fist ruled and only the physically toughest seemed to survive. Linny appeared to delight in wrestling me to the ground without warning, the way a king dog will bowl another and stand there growling just to let him know who's boss. I wasn't going to put up with this without a fight — I had a height and reach advantage — but I couldn't beat him that way. One evening, I was told to fill the Land Cruiser with petrol, and I made the mistake of leaving the ignition on after I'd checked the fuel gauge. When Linny went to start it up at 4.30 the following morning, he found that the battery was flat. My name was mud! He gave me a pair of the shears you use for clipping foot-rot from sheep's feet and told me to trim the hedge around the Okuku homestead — all 75 metres of it. It took all of a Saturday. But I soon found that by using my strength and enthusiasm in the right way — shouldering an 8-foot strainer post and marching with it straight up a 300-metre hillside, for example — I gradually earned his respect. I learned the ropes and adapted to this new way of doing things. I think I surprised a few old hands, who had openly reckoned I wouldn't last five minutes on the station.

The most important thing in ranking a shepherd or a musterer is the quality of their team of sheepdogs. Trouble is, you can't just go into a pet store and buy a team of sheepdogs — well, not without writing a very large cheque, anyway. You have to break in and train your own. And the other trouble is, good dogs are like hen's teeth. The rule of thumb in the high country is that a good sheepdog is roughly equal in value to a month's wages for a shepherd. Because a working dog only has a useful life of about seven years on steep hill-country work, it's hard enough for a well-equipped shepherd to have enough good dogs for their own needs without selling any off to new shepherds.

The traditional way around this is for a newbie shepherd to get a good old dog who is semi-retired but still up to light duties, such as yard work or mustering the killers (sheep that will be killed on the farm for dog food or mutton) from the flats. Ideally, this dog is kind enough to teach the young shepherd some stockmanship skills. Then it's up to the shepherd to try to get someone to sell him a well-broken-in dog that's young enough and obedient enough to be allowed to work on the hill (an unruly dog being worse than useless). From there, it's a matter of building up a team. To be considered an experienced shepherd, you need a full team, comprising (at the minimum) a good, strong-eyed heading dog — the type that can steer the mob in the right direction or bring a mob back to you through sheer force of character — as well as a couple of well broken-in huntaways, the type that will make a lot of noise and chase sheep or cattle away from them. Besides these, it's considered advisable to have a few young, up-and-coming pups that are under basic sit-down-and-come-back control, so that you can replace the older dogs in time.

With help from Linny and the head shepherd — a very keen dog-man named Maurice Flood — as well as senior shepherd Greg Spencer, I was able to gradually build up a semi-useful team. My first sheepdog was Bruce, an old black-and-tan huntaway who cost me $100 — about two weeks' wages. He came from the well-known dog triallist Kenny Barwell, who lived just down the road from Okuku at Loburn. Kenny was very good to young shepherds, helping them on their way by dealing in older dogs that were in semi-retirement. Bruce had a great bark and was useful in the yards, but his days of long runs out on the hill were over. He would happily amble along behind my horse, but as he could run only 50 metres at a time, I soon learned the importance of getting into position in a muster before making any noise. I once lined up with Bruce in the straight hunt at the Loburn Collie Club dog trials, alongside the famous high-country author and dog-man Peter Newton. Well, someone had to come last!

Bruce was soon joined by a young bitch named Sue, whom I got from the legendary Dick Carmichael, the manager of another Lands and Survey block, Coringa, about 50 miles away at Motunau. She cost me about a month's wages, and I was very lucky that Maurice Flood, himself a top dog-man, helped me break her in. Sue turned out to be almost the opposite to Bruce, keen to run out on the hill, and a bit hard to stop at times. But she had a very good nature and was eager to please.

Linny had worked on West Australian cattle stations, and had kept his hand in horsemanship by competing in rodeos. He was recognised throughout the land as a very good horseman, and he ran a little sideline business breaking in horses that no one else could seem to handle.

He had a winning formula with the difficult horses. He would break them in, and then ride them mustering in the steep hills to take the sting and buck out of them. Then he would hand them over to Maurice Flood or Greg Spencer, who would ride them vigorously for a few more weeks. After that, it was my turn. A horse was judged well broken-in when it couldn't buck me off anymore.

The system worked reasonably well until we came up against a little kid's pony that had earned a reputation for 'dirty tricks'. I was instructed to ride it out on the station and see that it had the sting taken out of it. Things were going reasonably well until I decided I could break a golden rule that Linny had drilled into me: never ride one of these horses downhill. They were always to be led.

After it had fooled me into a false sense of security, I let this dirty little pony gallop over the brow of a hill and onto the downhill slope, whereupon it seized its chance and worked a few very efficient bucks into its gait. When you fall off a largish horse, you normally get enough airtime to sort of roll into a bit of a ball and land more safely. But in this instance, being so close to the ground, there were no countermeasures available and I landed heavily on my back. It hurt like hell, but I wasn't about to let on to man or beast about this. Instead, I got straight back on to finish teaching the little sod a lesson. My lower back stayed very sore, but there was no use complaining.

'You'll live,' Linny would snap, if he was feeling inclined to mince his words.

He was implying I was just being soft, encouraging me to believe it was all in my head and I should just get on with it. And that's what I did, in spite of the pain. Every now and then, I had it investigated, but nothing showed up on the various scans and X-rays. It wasn't until I was about 30 when a very astute chiropractor thought to X-ray a different part of my spine altogether.

'You'd have been in a fair bit of pain for a while?' he guessed.

I agreed that I had.

'Well, that's not surprising. There's a break here,' he said, pointing to a spot in the middle of my spine, 'that's partly healed up. Don't imagine that's been too comfortable.'

The learning curve on Okuku Pass Station was as steep as the country, but it was an amazing experience. One of the most useful bits of knowledge I gleaned was from the policy of dividing the flock of 5000 ewes, with half being put up on the tussock high country on Mount Karetu, and the other half held on the lower country, which was rolling and improved. The tussock-block mob were to lamb on their own with no human intervention, while the other half were lambed by shepherds working from horse, motorbike or Land Rover.

When the time came to count the lambing percentage at tailing (or docking, if you live in the North Island) we learned that the ones up on the hill had done about 5 per cent better. It seemed counter-intuitive, but it showed that letting Mother Nature do her own thing without interference was actually better, and it cost a lot less. What's more, the theory was that it helped select for ewes that were good mothers. One of the consequences of assisting ewes that were in difficulty is that you preserved genes that didn't necessarily produce competent lambers.

Needless to say, as someone who had developed a faith in his own way of doing things, I butted up against Linny and the other head shepherds often enough. It wasn't until I was dealing with staff of my own that I began to understand how frustrated they must have been with me for trying to take shortcuts on a job!

The conviction you have when you're young that you're bulletproof got in the way sometimes, too. I got a royal telling-off one day when I spotted a cow stuck in a fence at the bottom of a very steep gully near the back of the station and took a tractor to pull her out. It was pretty hairy getting down, and even hairier reversing up and out, but I managed it and was very proud of myself. But if I was expecting congratulations from Linny, I was in for a disappointment.

'What the bloody hell did you think you were doing?' he raged, in words to that effect. 'You were 15 miles away over a mountain pass, no one knew you were there and you had no way of calling for help. What would have happened if something had gone wrong? You could have injured or killed yourself, or' — I swear he went pale at the thought — 'you could have lost the bloody tractor! You've got no idea about the paperwork I'd have to do if that happened.'

'I calculated the risk,' I replied. 'I decided I was safer on the tractor than on one of those bloody horses. If I'd got bucked off, I could have been dragged.'

'That,' said Linny, grinding his finger into my breastbone, 'is why we've got rules. You wear hobnail boots, not rubber, so you don't get caught in the stirrups. You've got a sheath knife and you keep it sharp enough to cut the stirrup leather off if you find yourself being dragged.'

There was no answer to this. These *were* Linny's rules. To some people, it might have seemed like Hobson's choice: a dangerous traverse of slippery ground on a machine or an equally dangerous expedition on an unruly horse. But what people — especially OSH regulators — don't understand is that experience is the thing you get the split-second after you actually need it. Without the freedom to put yourself at risk, you'll never acquire the experience necessary to form an accurate judgement of the best and safest way of getting things done. As a nation, we pride ourselves on being 'can do' kind of people. We're in danger of losing all that.

After a year on Okuku, it was time to move on to the other half of the practical component of my Dip Ag. Despite the Great Linno's iron rule, I left the station with a few regrets, the biggest of which was the need to let go of my team of dogs, which I had painstakingly worked into a high state of usefulness. They would be no use in the next round, which saw me back on a cropping farm, this time the 500-acre concern at Kirwee owned and run by Alastair and Liz Frizzell. I sold Sue to a mate, Jimmy Thacker, who was moving to work for James Innes at Haldon Station. I heard later that James Innes rather wanted her for himself, which pleased me no end. Having a good dog was a badge of honour amongst high-country musterers, and for me to have one of my dogs coveted by someone I had looked up to as a farming god since I was a young boy was a great compliment. Perhaps to soften the blow, later, when I turned 21, Michael Deans gave me a pup out of Alice, his Labrador bitch. I named him Dylan (the first in a long line of Labs with the names of musicians, such as Seager; my latest is Marley). He was to be my companion and partner in crime through all of my greatest adventures for the next nine years.

The Frizzells ran a piggery, and grew their own grain for pig food, so there were lots of different things to learn there. I learned to plough and grow crops, and a lot more besides. Pig farming, as Alastair readily admitted, was not the road to social advancement in Canterbury farming circles, but it was a great way to make money, particularly when you also grew the grain to feed them. He had an Agricultural Commerce degree from Lincoln, and he very soon had me constantly calculating the costs and benefits of everything that we did on the farm in my head. He also very generously allowed me to run a 'pen-to-pantry' boutique pork business, whereby I would take orders for a minimum of half a packaged pig from my Christchurch friends, kill the pigs to order, have them butchered, and deliver the meat to their freezer. This way, Alastair got to sell the pork at full market price and I got to subsidise my wages with a very useful retail mark-up!

One of the other real perks of the job with the Frizzells was that the farm had a very well-equipped workshop. In recent times, in addition to my ski-patrol activities up the hill on Mount Cheeseman, I had got very keen on 4WD rallying, a sport that differs from car rallying in that you are scored by how far you can drive through a difficult off-road section, such as deep bog or up a very slippery hill, sometimes against the stopwatch. In order to compete, I had bought Jumbuk, my Series 1 Land Rover, and I was able to use my spare time and Alastair's facilities to pull it apart and rebuild it with extensive roll-cages, bigger and more aggressive wheels and, of course, a 6-cylinder, 115hp Holden aboard instead of the original 4-cylinder, 55hp mill.

Just as I had always enjoyed trips to the skifield when the weather was against us, I had always relished adverse driving conditions. Ever since I'd been a young boy, I had loved the challenge of getting a tyre through a muddy puddle where no one else could. A family trip was always more exciting when the car got stuck — providing we got it out again, of course. This is what had drawn me to motocross. My career in that sport wasn't memorable for anything in the way of regular visits to the winner's podium, but it was notable for the fact that I made do with old, broken-down gear. It was an expensive sport, and while I was very keen on it, I didn't have much show of out-performing those who had the advantage of a big budget behind them. Besides having a very worn-out, very second-hand bike (which I bought with a small sum I inherited from my British grandmother), I had to pay for all of the maintenance and entry fees from my holiday earnings and repair work on school friends' cars. Steven Duff, a very successful rider who went on to have a large car dealership, sympathised with what I was trying to do and used to give me his half-worn tyres, which I would re-groove and sharpen with a hacksaw.

At the end of the day, I was a gangly, immature schoolboy trying to compete with 20-year-olds who were earning and fully grown. It was also a pretty physical sport and not for the faint-hearted. The skills learned in motocross were very handy when riding a farm bike quickly to muster wild cattle or head off a lamb break. But the fun of motocross wore thin quickly when the pain from falling off exceeded the excitement of flying past the checquered flag on one wheel or in mid-air.

As my enthusiasm for motocross waned, I dedicated myself instead to working out how my VW Beetle might climb a snowy skifield road better than everything else. And it was constantly butting up against the limitations of the Beetle that inspired me to buy the Land Rover. Of course, it wasn't enough just to be able to get through the deepest mud as well as any other Land Rover owner. I wanted to do it better than them.

The sport of 4WD rallying was in its infancy when I got on board. I attended meetings of the recently formed Christchurch

Quadrive 4WD Club, and hung on the every word of experienced rally drivers such as Hamish Wilson (who drove a 3-litre Vauxhall-powered Land Rover with widened mud tyres, roll-cage and hip-hugging bucket seats), as he regaled us with stories of the Deadwood Safari in Wellington, where he traversed impossible, tyre-filled mud holes, conquered near-vertical hill-climbs with names like 'The Widow-Maker' and sidled across hillsides that few drivers could avoid rolling off, all for the glory and the prize of a flash new American winch.

I was hooked.

It wasn't long before I became fully focused on getting my name recognised amongst the other great drivers. But of course, as we added more power from bigger engines — it wasn't long before I upgraded to a 186 Holden engine, to get as much power as I could without crossing the 3-litre threshold — and more grip from better tyres, something had to give. Broken axles and differentials were a constant handicap, although we soon learned to change a diff or axle while lying in the mud in about a quarter of the time it would take a Land Rover-certified garage to do it in a clean, dry workshop.

My first great success came during the Nelson 4WD rally held up the Maitai Valley just out of Nelson city. To this day I still remember the engine screaming at maximum revs as I threw all mechanical sympathy out the window (no, that's not right — we had removed all the windows to lighten the Landy as much as possible). My co-driver was Hugh Chapman, an old mate with whom I'd learned to drive on his family farm. Hughie held on with grim, silent determination as I flogged the Landy around a

1-kilometre speed section through bogs, over small gorse bushes (something Hughie and I had practised about 15 years before when we were four-year-olds driving an old Chev truck while his father fed out!) and up shingle banks, to slide, only just under control, to a garage stop. Four pegs emulating the size of a one-car garage had been added to the obstacles as a nod in the direction, perhaps ironically, of discipline and tidy driving. The timekeeper kept looking at his stopwatch in disbelief, and eventually asked us if we had taken a shortcut. He and we all knew that was improbable, because in the vicinity of the top corner of the course the gorse was 5-foot tall. Driving through it rather than on the rough track would have slowed us down, if not bellied the Landy out.

It wasn't until the prizegiving at the Rutherford Hotel that night, when I was called up to receive a silver cup and an armful of product prizes for being the fastest vehicle in the speed section, that I really believed we had done it.

Such success can go to your head. When I was competing in the Nelson rally the following year with a reputation to keep up, I had possibly the biggest fright I have ever received in rallying. It was a different course, and the speed section part was much more open and therefore faster. I had driven it in practice, but at a sensible pace.

In the heat of competition, I am slightly ashamed to admit that the red mist descended over my better judgement and, as I plunged down into a valley at full speed towards a narrow bridge, I realised with a sick lurch that the angle and speed of approach were not going to line up with the capacity of my brakes to slow me down in such a way as to allow a controlled crossing of the bridge. In a split-second, I realised I had two options. The more sensible would have been to aim the Land Rover up the bank before the bridge to scrub some speed, so that the approach to the bridge could be made in a relatively safe and controlled manner. The other option was to cut the corner and accept that the final few metres of the approach would probably be airborne and that, with the wheels off the ground, there would be no point in applying the brakes. This option critically depended upon the 8 x 3 timbers on each side of the deck of the bridge preventing the tyres going over the edge and sending us plummeting into the creek, 10 metres below.

In the event, the Landy decided for me. The brake drums, probably red-hot by now, had expanded away from the shoes, producing what it is known as brake fade. There was no way I could regain sufficient control to execute option one.

There was a sudden, breathless halt in the rattling, swishing and banging of our progress as we sailed through the air. Then, with a crash, we landed on the slick timber of the bridge and slithered to and fro for an instant before I could apply the power and get all four wheels in line again. After that, my focus reverted to achieving maximum speed up the other side of the valley and towards the finish line.

I don't recall the prize that time, as, like your second cold beer, it's never as good as the first. But although it was nearly four decades ago, I remember the drive with utter clarity. I had nightmares about that episode for many years afterwards, and it's always at the back of my mind when I'm instructing 4WD drivers in the art of off-roading and advising them not to rely on their brakes too much in low-traction situations. Hughie and I have never discussed it — not then, and never since. Perhaps we should. I don't know whether he had made the same mental calculations as I had, and fully realised the possible outcomes. If he had, I'm sure he's as haunted as I am.

Sorry, Hughie, if I gave you a bit of a fright.

At the end of my year at Kirwee, I took a summer holiday job as a general hand on my family's farm, Waipari Station up in Hawke's Bay. It had been in our hands since it was carved from the larger Mangakuri Block in about 1900, but because the family had always been involved in ecclesiastical duties elsewhere — like Dad after him, my Grandfather Alwyn was a vicar and then a bishop — the Warrens were absentee farmers, and Waipari was under a manager. After that, it was time for the theory-based book-learning component of the diploma. The lecture halls of Lincoln weren't my natural habitat, especially after having spent so long free and unconfined. But Dad was adamant that if I wanted to become a farmer and, more particularly, if I harboured any ambitions of taking on the family farm, there was no other way. I resigned myself to a year of pain and a significant drop in my income.

I'm not sure I fully appreciated the value of what I was learning at the time: that would come later. As for the loss of income, thanks to my considerable expertise with Land Rovers (there's nothing like off-road rallying on a budget to teach you how to diagnose and fix mechanical problems), I was able to partly offset that by getting a job working in a truck-wrecking yard. This came about after one of the many trips I made from Kirwee to Christchurch, on the prowl for parts for the Landy. On this particular occasion, I visited Autorama on Blenheim Road in the industrial hinterland of Christchurch. It comprised a front yard with some very second-hand trucks for sale, a large parts shed, grimy offices and a yard filled with a couple of acres' worth of broken, wrecked trucks, paintwork faded and rusted, window glass shattered or glazed with dirt — promising rich pickings for such as me. Sure enough, I soon discovered a pile of old Land Rovers that were just waiting to donate parts to mine and be a part of some adventures again. Repowering my Landy with the Holden had subjected the drivetrain to a bunch of stresses and strains that its designers had never contemplated. Consequently, when I found a large stockpile of 4WD transmission parts, I realised I was sitting on a bit of a goldmine, to say nothing of a competitive edge. My excitement must have been visible, as the owner asked me if I would like a job selling them. I told him I already had a good job at Kirwee, but mentioned that if the offer were still open when I was at Lincoln the following year, I would definitely be in the market for a part-time job. I was hired on the spot — and given a generous discount on the diff I had bought. And when I presented myself back at Autorama after my first couple of days settling in at Lincoln, I received a warm welcome.

I worked pretty much every Saturday. The art of trading in used mechanical gear is being able to see past the rust, the grease and the dirt to the quality of the steel, and for someone who had a few clues mechanically and who could do their sums quickly, what appeared to the untrained eye to be a sea of crap was actually a valuable resource. I was good at the job. It didn't take long before I was promoted — very informally, of course — to weekend manager, and I soon developed some lucrative sidelines to the main business as well. Having access to low-cost spares enabled me to perform a few budget repairs to my classmates' vehicles. I was able to help out other Land Rover owners when I attended events of the Christchurch Land Rover Owners' Club, and my 4WD mates soon cottoned onto the fact that I was always flush with parts, and it wasn't long before I was being asked to source this and that. By that stage, I was quite committed to the whole 4WD rally scene, and most weekends I would trailer or A-frame the Landy (which wasn't quite road-legal) to meets as far afield as Nelson or Dunedin. I'd usually throw in a few spares on the off-chance someone would break a diff or axle and need a speedy replacement. The owner of Autorama was very generous in calculating my staff buying price, which meant I got everything for not much above cost price. Being in the position to supply a scarce commodity at a time of high demand worked well in my favour. I would collect a 'fully retail' price in cash, which I'd present to the yard owner on Monday, less my commission. He would happily peel a $20 note or two off the roll and hand them back as a bonus. Modesty prevents me from admitting my sideline profit that year but, suffice to say, I finished my year as a struggling student with no debt and having added a late-model Chrysler Valiant — a 6-cylinder, luxury tow-car for lugging the

Landy — to my vehicle fleet. And quite apart from the money, the wealth of experience I gained at Autorama stood me in very good stead in my future adventures.

When I finished my diploma in 1981, I took a job, as so many of the students do, as a truck driver for the Wattie's pea harvest. This involved shift-work, starting at 2pm and often finishing around 2am, driving a Ford D900 12-ton tip-truck laden with about 8 tons of peas fresh from the harvester in the paddock to the factory. As the pea season drew to a close, I got a job as a trainee truck and diesel mechanic with Ian Giltrap Ltd in Christchurch. The job entailed repairing and building tip-trucks and tractors, as well as recovering broken-down machinery from used-machinery auctions around the country. Ian seemed to appreciate my ability to think outside the square, and the experience I had gained in vehicle recovery from 4WD rallying was put to good use. I didn't complete a formal apprenticeship, but the 18 months I spent at Giltrap's turned me into a pretty resourceful mechanic.

The job wasn't without incident. There were times that, when recovering machinery, I may have pushed the loading limits a bit. On one trip, I was bringing a digger back from Wellington on an articulated Dodge K950 truck. The digger had a leaking main-boom ram, which meant that if the boom was slung to the side it might settle down and wipe out an oncoming vehicle. I had to stop every so often, start the machine up and lift the boom as high as possible to stop it flopping over sideways and into the path of oncoming traffic. It turned out that with the boom hoisted high like that, I was ideally configured for determining whether the tieback wires steadying the power poles had sagged, because, as I was rounding one of the tight cliff-side bends north of Kaikoura, there was a sudden snagging feeling and a few hefty bangs and groans as the digger boom hooked up on a cable and ripped the entire pole assembly clean out of the ground.

Imagining I might have collected some power lines and that they might still be live, I brought the truck to a halt 100 metres on from the site of the incident, trailing 50 metres of lines behind me. Then I jumped well clear of the truck, in case it was sitting there live on its rubber tyres, and did the only decent thing — cut the wires and left the pole on the side of the road.

Naturally, when up to 500 toll calls are rudely interrupted all at once, the boys in blue tend to take a keen interest. Just out of Kaikoura, I saw a police roadblock set up — four cars, lots of policemen and traffic cones — all in my honour. Of course, conscious of my legal responsibilities, I had taken care before setting off again to ensure that, even with the boom hoisted, the total height of my load was slightly under the 4.25-metre limit. After inspection, the police waved me on without further ado.

I got back to the yard, pleased with my week's recovery efforts, and I had no doubts Ian's keen eye for a bargain would have ensured the digger would fetch a tidy profit down the track.

Ian was standing in the yard waiting as I drove up.

'How did you get on?' he asked.

'Very well,' I replied. 'Apart from catching the tieback on the main phone lines north of Kaikoura.'

'How high were you?'

'Under 4.25 metres.'

'No problem,' said Ian. 'So long as your height was below 4.25 metres, you're allowed to take the lines with you!'

Nothing more was ever said.

Not only was the Giltrap job good training, but I was able to purpose-build a Toyota Land Cruiser rally vehicle in their well-equipped workshop. This involved stripping the vehicle to its bare chassis and building a full roll-cage, fitting re-tuned suspension and rebuilding and hotting up the motor under the direction of my mate Pete Cook, a mechanic just back from building V8 Supercars in Bathurst, Australia. Pete's ability to make engines go as fast as possible is legendary in Canterbury, and by stripping away all unnecessary metal and bolting on a few choice additions, he soon turned my motor from a lazy truck engine into a livelier racing engine. For the technically minded, it had 60 thou shaved from the head to increase the compression ratio, a port and polish to enable the air/petrol mixture to flow more quickly into the cylinders, balance and blueprint to enable it to rev more freely, a racing jetboat cam grind to increase more low-down torque, and was fitted with a Holley 350 carburettor to get a greater fuel-flow into the engine and a very open, big bore extractor exhaust to allow all the spent gases to exit as quickly as possible.

He must have done a fine job, as not very many years later we had won (twice in a row) the V8 production (modified) class

of the New Zealand 4WD Association national rally series —
and we only had six cylinders! However, with the boundaries
we were pushing with our modifications, it was getting
increasingly hard to argue that it was a standard Land Cruiser
motor anymore. It wasn't exactly normal for the motor of a
Land Cruiser, designed as a hill-country cocky's workhorse, to
develop over 200hp at 7000rpm (unmodified, it would normally
only rev to about 4000rpm, producing 155hp), and nor would
the most enthusiastic farmer expect their Land Cruiser to exceed
100 miles per hour (mine managed 160 kilometres an hour). It
was little surprise we were 'moved out' of the production class.

The medium-term plan (it was actually my short-term plan, but
Dad had other ideas) was that I would take over the management
of Waipari Station. After 18 months being a grease monkey,
I was about ready to get back amongst the hills and get back
to my farming career, so I moved up to Waipari and resumed
work there as a junior shepherd under the manager. But I soon
became frustrated with the way I saw things being done. When
I felt I wasn't learning enough, I decided I would be better off
working for a farmer whose ideas more closely aligned with
my own. When I was offered a job by Les and Sue McHardy,
who farmed 20 miles south of Waipari on Te Aratoi Station,
I jumped at it. Te Aratoi is situated in the steep hill country above
Blackhead Beach on the coast east of Waipukurau, famous in the
early twentieth century for the wreck of the schooner *Maroro*,

which occurred there in 1927. The wreck was a local landmark for many years.

Les and Sue ran a very efficient farming operation. Working for them, I was able to build up a new team of sheepdogs, and I learned a lot from Les about stockmanship in the Hawke's Bay — how to read livestock and look for a 'bloom' in the lambs' wool, which meant they were healthy and growing well, and to their children apportion the correct grass to the neediest or most responsive animals. Les and Sue looked after me very well, warmly accepting me into their family. I very much enjoyed being babysitter to their children, Alex, Nicola, Douglas and Sarah, who are all grown-up now and making their way well in the world.

The comparative isolation of Te Aratoi didn't bother me too much. My sister, Celia, was living in the singlemen's quarters on Waipari, only half an hour's drive away. She was running a riding school, which meant that there was lots of social activity with her riding friends on offer. Mostly, this comprised woolshed parties around the district. Woolshed parties are a bit unique. The wool-room floor, the boards worn slick by the dragging of wool bales across them and polished with lanolin, makes for an excellent (if slippery) dancefloor, and the bar can be set up out the back on the grating of the catching pens, slatted to let sheep droppings rattle through, where a bit of spilled booze is no problem whatsoever. Similarly, with a wool fadge slung up in a corner, the slats serve very nicely as a unisex urinal.

I was enjoying life at Te Aratoi, but nevertheless had started planning to do the done thing and head off overseas on my OE.

But one day in late February 1984, just when I was sitting down to lunch, the phone rang. It was Dad.

'We had dinner with Mark Williams yesterday,' he said.

'Oh, yeah,' I replied. Mark was a distant relative who farmed a property neighbouring Waipari, and such was Dad's dissatisfaction with the way our family block was being run that he had hired Mark as a supervisor.

'There are big changes. We're going to invite the manager to move on. You're going to take over the management on July first.'

I was momentarily speechless.

'But I was planning on going on my big OE!' I exclaimed.

'Better make it snappy, then,' Dad said. 'We'll expect you back on the thirtieth of June at the latest.'

'I won't let you down, Dad,' I replied, feeling more lightheaded than anything else.

I broke the news to Les. I told him I was sorry to be leaving him, and I meant it. He and Sue were wonderful — to all appearances more pleased for me than anything else. I suppose it had always been obvious that it was my dream to be the first Warren to be the hands-on farmer of the family property. They helped me book flights and get my act together, and shortly before I was due to leave Sue even made me a sweatshirt emblazoned with the words *I am proud to be a farmer*, to wear instead of my thick, woollen tussock-jumping jersey.

I had a few conversations with Mark Williams. It wasn't altogether a surprise that he and Dad had decided to re-plan the management, as it was pretty plain to everyone that the farm was going backwards. Mark had advised that he felt it would be hard to attract another manager to take on the enormous task of turning the place around, and that's how I came to be selected for the job, three years before I had anticipated it. Mark was confident I could make a go of it, under his guidance. 'Hopefully he'll only make mistakes once!' he said to Dad.

The other farm staff all went down the road as well. One job I did manage to do before setting off overseas was to hire a fencer-general. Maui ('Hoke') Hokamau was working on a neighbouring station, but heard the news about the regime change at Waipari on the bush telegraph. He phoned and expressed a desire to come and work for me.

'It'll be a bloody big job,' I warned him.

Hoke had seen the scale of it, but he wasn't fazed.

'I reckon I'll be right for that,' he said.

I was delighted. Hoke was a highly competent fencer. He went on to win many fencing competitions at the National Fieldays. We at Waipari can possibly take some credit for that: we gave him plenty of practice!

When the day came, Les and Sue delivered me to the airport. Les crushed my hand in the firm grip of his own.

'All the best, mate,' he said. 'You'll do fine.'

But it wasn't his reassurances that were ringing in my ears as the plane gathered speed along the runway. It was my Granddad Alwyn's words.

'The farm's in trouble, Mark,' he said. 'If you fail, you've lost the family fortune.'

No pressure, then.

CHAPTER 5

THE FIRST OF A MILLION STEPS

I sighed.

There was no way I was going to get a satisfactory bath in my Tokyo hotel room. The tub didn't seem to be any good as anything but an outsized dog's water bowl. I decided I'd be better off trying to get some sleep before the 13-hour flight back to New Zealand. It had been a great three month whirlwind trip through Europe and Japan, but it was hard to keep my mind from straying to what awaited me when I got back.

Sleep was elusive. The bed was no better than the bath, clearly designed for guests built on a far more modest scale than a 24-year-old, 6-foot Kiwi farmboy.

I got up and tuned the TV into an English-speaking news station, and was mildly shocked to see an item about New Zealand. Ten days before, the prime minister of the National

Government, Robert Muldoon, had called a snap election. It was probably a good thing I didn't have a crystal ball then, because if I'd known how much political developments were about to complicate the job ahead of me, I'd probably have enrolled in Japanese lessons on the spot.

I arrived in Napier on the afternoon of 27 June, a bit jetlagged and out of condition from good eating and a three-month break from physical work. Celia was there to meet me, as was, waiting out on the Waipari Land Cruiser in the car park, Dylan, my golden Labrador. It was a real thrill to see him on the Cruiser — he was my best mate — and to realise that the battered old vehicle was now my responsibility, along with the station itself.

When the Mangakuri Block had been carved up in about 1900, the grand old homestead had gone with the Clareinch Block over the road. This was the source of some regret to my family, but perhaps we were fortunate. Over the years, I have noticed how the cost of maintaining a grand homestead often drags down the farm supporting it, and, in fact, the Clareinch homestead was demolished a few years ago because the ongoing maintenance bill was more than the farm could afford.

Waipari, by contrast, was run from a residence that was built for its first manager, Ernest Gilbertson, in 1930. Every expense was spared at the time — it cost a measly £700 — and it was definitely built for economy, not comfort. It was a three-room affair with matai floors but fibrolite walls, and completely without insulation. Apart from the odd extension and modification made ad hoc to accommodate the managers' growing families, nothing much had been done to it in all that time. It was adequate — just. On my first night there I found it cold, and with a constant draught blowing through the empty rooms. Beyond a kitchen table and the odd set of drawers, there was no furniture. Celia had thoughtfully made up a bed for me on the floor.

The first thing I did after dumping my bag was pick up the receiver and crank the handle to deliver two long rings on the party line, the signal for the bullock cottage where Hoke and his wife, Sandy, had set up house.

'Welcome to Waipari,' I said, feeling decidedly odd. It was my first-ever call as a farm manager to my first-ever staff member.

'Well, welcome to you!' Hoke replied. 'I got here first!'

'What time have you been starting in the morning?' I asked.

'Oh, around eight,' he replied.

I was surprised. On most of the farms where I'd worked over the years, we'd have already put in two or three hours by then. But Mark Williams and others had warned me not to go too hard, too early.

'Eight it is,' I said.

The job had begun!

Waipari comprises about 3200 acres (1300 hectares) of mostly class 6 country, which means it is pretty steep. For all that, though, it lies quite well to the sun and, being coastal, the winter is not as harsh as it is for properties further inland. When I took over, things were being run much as they had been since the early days. Because of the steepness of the country, most of the stockwork was also done as it had been since Europeans first settled the land, on foot or from horseback. There were horses everywhere, a few of the bloodlines reflecting the long-standing obsession Mum and Celia both had with horses, from Celia's pony-club days to her more recent attempt to run a riding school. Some were useful station hacks. Others were not so much.

We were a mixed sheep and beef outfit, with the emphasis on growing wool and making a bit on the side from lambs, most of which were sold in the spring as store lambs. The cattle were mostly run as workers, keeping the grass down and in optimum condition for the lambs, but we sold the weaners as store animals, too. Strictly according to tradition, we shore our sheep twice a year.

And never mind what we were doing: the business model for New Zealand farmers had hardly changed since the first cargo of frozen mutton arrived in England in reasonably good nick aboard the *Dunedin* in 1882. Wool was still king, as it fetched good prices in international markets. Farmers raised their lambs to be as fat as possible and sold them to their local meat works, which killed them, dressed the carcasses, wrapped them in stockinette and froze them for export, mostly to England. The only significant innovation there had been in 100 years was the advent of the aerial application of superphosphate, which had dramatically improved huge swathes of otherwise marginal farmland.

My first day as manager was spent on a tour of inspection. It was pretty grim. Of the 45 paddocks comprising the 3200-odd acres, only about three were stock-proof. Ancient strainers were leaning at drunken angles, battens were missing and in places whole sections were lying on the ground. Gates were broken or sagging so badly they wouldn't close. Stock had been leaking to neighbouring properties at a furious rate, and in one case they weren't always being returned.

'You'll be busy,' I told Hoke, wondering if he had realised just how bad things were.

'I've already been busy,' he said. 'Can't you tell?'

He had been working to fix the worst bits, which was a big enough job, but going forward we needed to be strategic. We made a plan to ring-fence the property, securing the boundaries first and then concentrating on fencing the place in half and then in half again, so we could start to farm it properly. We also separated the stock so that we didn't have to redraft every five minutes to sort the ewes from the rams, lambs and killers.

The property was overstocked — a product of government policies that encouraged farmers to test their carrying capacity to the limit — and the stock were in very poor condition. The reason was plain to see all around us. The pasture was sere and patchy and studded with weeds — it had been a dry summer, but with little or no fertiliser having been applied in the past three years, and with the fencing inadequate to manage the grazing, the grass had failed to come back in the autumn.

When I sat down to inspect the farm diaries and the account books with our accountant, Pita Alexander, I found things were no better in the office. In fact, given the capital-intensive work of bringing the place back that confronted us, the work ahead bringing the property up to scratch seemed the least of our worries. The previous year's lambing percentage had come in at 56 per cent (that is, just over half of the ewes had produced a live lamb) and calving at 60 per cent. Even allowing for the hard season, the numbers ought to have been half as high again. Farm working expenses were at 99 per cent of gross income (they ought to have been around 50 per cent), and debt was running at 50 per cent of assets. While this doesn't sound too bad, it was the trend that was significant. Most of this debt had been racked up quite recently, and interest rates were running at 12 per cent on term debt and 14 per cent on current account overdraft. If you plotted the station's financial position, there was no mistaking the trajectory. It was nosediving. So what was to be done? There were no quick fixes on the income side of the ledger. So quite apart from getting the place running properly again by upgrading the fencing and improving the pasture in order to maximise income, we also needed to address costs. There are usually three big items in the outgoings column for a farming operation. One is the interest bill at the bank. Another is the cost of labour. The third is fertiliser.

We didn't have much scope to address interest rates. My eye was caught by a hire-purchase bill we were paying, on a brand-new 55hp John Deere 4WD tractor with front-end loader that the previous manager had bought on tick two years before. Hire purchase is a bit of a trap for farmers, as Pita pointed out. It's usually quite easy to get finance for machinery, as it can be secured against your land. But Pita reckoned it was often a symptom of a farm in trouble, as it signalled a lack of cashflow. We were paying 18 per cent interest, and it seemed obvious to me that we ought to shift the debt to our current account and save 4 per cent per annum. Trouble was (as it seemed at the time), the finance contract was for five years, and there were stiff penalties if it was broken. I had to grit my teeth and keep making the repayments.

By contrast with the shiny new tractor, the rest of the machinery was pretty clapped-out. The Land Cruiser was a 1974 model, and had 190,000 kilometres on the clock when the station bought it very second-hand from the Hawke's Bay Catchment Board. Very few of those kilometres would have been easy, and it wasn't exactly living in retirement on Waipari. I wasn't too intimidated by the prospect of making repairs every now and then in order to keep it running. Similarly, there were three pretty much buggered farm bikes in the shed. When Mark Williams asked what my intentions were with regard to these, I told him I would happily do the work necessary to keep them running.

'Come off it. You've got better things to do,' he said, shaking his head. 'You're a farmer now, not a mechanic.'

I reluctantly saw his point. We would need some new farm bikes.

Mark had sorted another set of expenses at one stroke by letting the manager and his considerable entourage of staff go. I had yet to determine the ideal number of hands needed to run the place efficiently, but the previous staffing level had been way over the top. In fact, there had been a joke going around the district that there were so many shepherds on Waipari that at docking time all they needed to do to make a portable sheep yard was hold hands! God knows what they'd been doing half the time. There was evidence everywhere of work that had been started, sometimes years before, and not completed. Every night after work for the first month, I would go out and swing another gate. The gates were actually there — lying on the ground with grass going through them, more often than not surrounded by the hardware (hinges and latches), other fencing materials and even tools. My delight at finding these little treasure troves of fencing gear in the grass was usually tempered by the fact that whenever I went to snatch up a box of staples in triumph, I'd discover the bottom had rotted out, and would be faced with the frustrating job of picking the staples up from the ground. But it was good, given the state of the finances, to scavenge these valuable materials. I repaired hinges and latches and straightened posts, so the gates would be stock-proof and easily opened from horseback.

I wasn't winning any popularity contests when it came to rationalising the equine side of our operation. Managing bloodlines by putting everything out in the same paddock and seeing what happened hadn't produced uniform excellence amongst our animals, and many of them weren't good for much beyond eating grass and looking like extras from a cowboy movie. But it wasn't easy getting Mother or Celia to admit this. I kept two or three decent station hacks. The better animals were given away or, in a few cases, returned to the South Island where Mother wanted them. As for the rest, well, let's just say the Dannevirke and Hawke's Bay Hunts had some contented hounds there for a while.

One of the first things I did upon taking over was a SWOT (Strengths, Weaknesses, Opportunities and Threats) analysis, in the course of which I identified the more extensive employment of Mother Nature as a distinct opportunity. After all, she worked for free and had thousands of years of experience. My experience on Okuku Pass had converted me to the notion of natural lambing, and according to this philosophy the big gang of shepherds who had previously done the 'lambing beat' in late winter were surplus to requirements. In the old days, we would all dress up in wet-weather gear and set about painstakingly touring the lambing paddocks by motorbike or on horseback. The idea was to help any ewe that was having trouble lambing by catching her, pushing an invariably less-than-sterile hand up her vagina and pulling out an oversized lamb through an undersized opening or, failing that, extracting a dead lamb bit by bit. All this really achieved, even if we succeeded (and often we caused more problems than we solved in saving the ewe and perhaps the lamb), was that we selected for ewes that had trouble lambing naturally. With fuel and labour costs skyrocketing, it was hard to justify the lambing beat. That was a whole lot of outlay on labour saved, right there.

Once I had settled in, I began interviewing for a permanent head shepherd. The bloke I eventually took on wanted to be known as 'stock manager', and so far as I was concerned he could call himself whatever he liked so long as he was a good stockman and prepared to work. He was hired on a salary of $14,000 per annum. Technically, I too was an employee of the family farm partnership, and I was managing what was, potentially, a multi-million-dollar business. But my salary was just $12,000 per annum and, given the state of the finances, Dad and Granddad Alwyn had forbidden me to draw more than $500 a month. To all intents and purposes, I was working for the love of it.

If that had been all we had to face, it wouldn't have been so bad. But things were about to get much, much worse.

While I was totally focused on getting the farming business back up and running, interviewing more staff and getting my head around the multitude of problems and the scale of the challenge, the rest of New Zealand was gearing up to go to the polls on 14 July 1984 for the snap election called by Rob Muldoon. As a nation, New Zealand had been living beyond its means ever since Britain — formerly the next best thing to a captive market for our primary produce — had entered the European Economic Community in 1972. The realities weren't helped by the two 'oil shocks' of the 1970s, which produced rapid inflation. Muldoon had attempted to keep a lid on all this with a freeze on wages and prices, but something had to give.

Quite apart from the fundamentals underpinning it all, and whether it was obvious to anyone at the time, New Zealand had tired of the level of State control typified by the style of economics practised since the Great Depression of the 1930s, whereby practically every aspect of the economy was centrally planned by the government. Heavily regulated, protected with tariffs and import rationing, Muldoon's New Zealand was known as 'Fortress New Zealand'. But by 1984 his administration was under increasing pressure from a set of major economic development projects collectively known as 'Think Big', which had entailed massive offshore borrowing. And to make matters worse, he had resorted to shameless election bribes, seeking to shore up his constituency by offering an unsustainable superannuation scheme to the old and an equally unrealistic set of subsidies to farmers. Crippled by the national debt burden, Muldoon had called an early election as a last, desperate gambit to cling onto power, hoping to catch the Labour Opposition on the hop. But with National's vote split by the New Zealand Party, a political party founded for that very purpose by businessman and property magnate Bob Jones, the gamble failed dismally. National got only 36 per cent of the vote and Labour swept to power with 43 per cent.

Few could have predicted what would happen next. When Muldoon opened the books to the incoming government with a show of vindictive glee, the new Prime Minister, David Lange, found himself confronted with a full-scale economic crisis. Within two days of the election, Labour had devalued the New Zealand dollar by 20 per cent. Interest rates soared overnight. Our current account with the Bank of New Zealand was suddenly attracting 23 per cent interest — and I was quietly delighted that I had been prevented from whacking the beehive on the hire-purchase agreement for the tractor. We were rather luckier than most with our term debt, which was with the insurance company AMP. Since we all held substantial life policies with them, they offered us a concession on their retail rate. We heard that some of our farming friends were paying up to 23 per cent on their term debt, and as much as 29 per cent on their current account. And needless to say, with the currency devalued by 20 per cent, the price of everything went up. Petrol shot up from the price at which it had been frozen since 1981, $3.20 per gallon (4.5 litres), to about $4 per gallon. Our produce was theoretically worth 20 per cent more in our export markets, but it didn't always work out that way.

I think I probably used some colourful language when I heard about the petrol prices — I'd already worked out the Land Cruiser had a drinking problem, getting about 13 miles to the gallon (even though I later converted it to run on LPG for half the price). But my focus at that stage was solely on getting the farm up and running. I was working towards my first docking in late spring. Because we'd muster in all of our lambs, we'd get our first opportunity to count them and see what return we were getting for our efforts. So significant was that number in the scheme of things that I don't have to consult my diaries to remember it. It's etched in my memory. We docked 4948 lambs from 6000 ewes, an overall percentage of 82 per cent. Although the number of lambs was roughly the same as the previous year, this represented a big improvement, because Waipari had been carrying far more ewes — 8000 — when I first arrived. (Although it was hard to get a handle on the precise number, because we found a fair few mixed-age wethers — castrated rams — marked as ewes.) Our lambs were in better condition, too, because the previous year's crop had suffered from both drought and overstocking. Some of those lambs fetched only $1 per head.

Because there was another dry season in prospect, I decided I would sell as many of my new lambs as practical. In those days, the Meat Board (which was the sole buyer of New Zealand livestock) would announce what was known as the alpha schedule, setting out the prices that would be paid for lambs for the early season, on 1 October. The schedule was divided into three grades. For lambs of 9 to 11 kilograms, they were paying $2.77 gross per kilo. For lambs between 11 and 13 kilograms, the return was $2.09 per kilo. There was a third grade, but since producing a 15-kilogram lamb was only a distant dream for us at the time, I don't remember the price. In the end, I sent my first truckload away for an average of $20 per head. I was quite pleased — and even more so when Jim Aitken, a neighbour whose opinion I valued highly, congratulated me and told me I had done well.

As the hours of daylight lengthened, so did my working days. I would typically be up at 4.30am to move stock in the cool of the day. I would either have a sandwich and a bit of keenly anticipated fruitcake on the hill at around midday, or occasionally allow myself the luxury of heading home for a bite to eat, where I would use the time wisely to catch up with the orders of farm supplies from town and incoming messages. Hoke's wife, Sandy, monitored the party line and acted as a secretary of sorts. Sometimes, I would have a quick snooze on my new (or new-to-me, anyway) sofa-bed, listening to the dulcet tones of Heugh Chappell reading the rural report on *Midday Report* on National Radio. When the tone of the broadcast changed with the *Maori News* coming on at the end of the rural bulletin, it was back-to-work time, finding jobs to do during the heat of the day that didn't involve moving stock. I would do more stockwork as the afternoon cooled and until darkness chased me home. It was an exhausting routine. One time Celia came up to the homestead to find Dylan and me asleep on the gravel in the shade of the Land Cruiser.

I was too busy trying to get a lungful of air to give any thought to draining the swamp. Consequently, it wasn't until the day after the revolution that I got my first inkling of it.

I had a TV — an old black-and-white set that someone had given me — but I never watched it. Partly this was because I was too tired or busy. Partly it was because it was a bit of hassle. In those days, we received a signal via a repeater station on Top Panama that was powered by a 12-volt car battery recharged by a solar panel. The battery was always running out in the winter, and constantly needed replacing and charging. When there was an All Blacks test match in the offing, there would be a co-ordinated effort amongst the locals to ensure there was a freshly charged 12-volt battery in order for everyone to receive two or three hours' uninterrupted viewing.

For these reasons, the Fourth Labour Government's first Budget passed me by. In those days, Budgets were announced live on evening TV. People used to make a bit of an occasion of sitting around a TV set to find out who were the winners and losers. There would have been some pretty sombre gatherings in the New Zealand countryside on that black evening in 1984, when farmers learned from the Minister of Finance, Roger Douglas, that all government assistance to industry was to cease, and that agriculture could not hope to be treated differently to any other industry. As it turned out, and probably because farmers weren't known for voting Labour anyway, agriculture was hit first and hardest by the sudden change in economic climate.

To understand just how rapid and savage the change was, you need to have a sense of how entirely New Zealand farmers depended upon subsidies of one sort or another. At the time a 13-kilogram lamb was worth just over $20, but a year later with subsidies removed it was worth not much more than half that.

Meetings were hastily arranged in rural halls all around the country. Reactions ranged from shock and disbelief to profound pessimism. For many, it was like hearing you'd been given a date with the firing squad. I recall a farming mate of mine, Peter Wynn-Lewis, standing up and telling a tense meeting in Waipukurau that he had just bought his farm the year before. He had cut costs wherever he could — he was shearing his own sheep with his new-born daughter in a cot on the shearing board while his wife, Jo, went teaching to pay their grocery bills — but now, with interest rates at 23 per cent and likely to rise, he was certain he would go to the wall. True to his word, after struggling on gamely for a while, Pete and Jo eventually shifted off to town, where he felt he could use his Agricultural Commerce degree to much better effect.

Our own business had been marginal *before* the subsidies were removed; I wondered if I might not be better off doing what Pete had done.

***

I had to make a go of it. Dad was supportive but, as he lived in Hornby and had never actually farmed Waipari, he was of limited use. I relied heavily on the collective wisdom — so generously offered — of friends and neighbours. Mark Williams on the back boundary was great. So, too, was Neil Kittow, who farmed just over the next ridge. On the strength of the 'economic farm surplus' (or EFS, as a farmer's profit was known in Ag Com circles) he had consistently achieved, Neil had won the only Farmer of the Decade award ever presented. He was completely open with me about his farm figures and methodology, quite apart from the wisdom of the advice he offered when I was feeling the pressure of it all. Others, such as Neil and Les McHardy, Peter Nancarrow, Hugh Williams and Bay de Lautour, were also great listeners and advisers, being very patient with my (sometimes unusual) questions.

I could see no way through but to work harder, so that's what I did. I insisted that my staff did, too. This didn't necessarily go down well with all of the blokes I took on. To start with, when they muttered about the long list of jobs I was dishing out each day, I tried explaining the economics to them, to help them appreciate the urgency and have some input into the decision-making process. In most cases, I shouldn't have bothered. When I tried to show them the workings of the cost-benefit analysis I'd done on a certain practice, for example, or attempted to talk them through the terrible logic of compound interest by explaining the 'rule of 72' (divide 72 by the interest rate and you get a figure representing the number of years it takes for compounding debt to double), they'd throw their hands in the air.

'Jeez, boss! Spare me! As long as I've got enough dough for smokes and a beer at the end of the week, I'm happy!'

It was tough getting some of them to view farming as a business. Most were inclined to think that riding about on a horse looking busy was a fair day's work for a fair day's pay, and if it rained, why, you took the day off and headed into town. I couldn't afford to indulge these attitudes. I let a few hands go, and not all of them were happy about it. Perhaps I could have been a little gentler in my manner. Pita Alexander had told me that a good rule to live by in the business world was 'Be ruthless with money but gracious with people.' I might have got that mixed up at times. I soon gathered I was getting a reputation for being a bit of a demanding young upstart. But I was heartened when one day I heard that the usual bitching session had been underway at the local pub with my name mentioned in dispatches, only this time my stock manager had been within earshot.

'Why would anyone want to work for that bastard?' someone was saying, sucking sourly on his beer.

'I work for him,' the stock manager piped up. 'He's OK to work for. He just doesn't put up with fuckwits!'

In my second season, and following another dry summer, I again judged it prudent to de-stock. I had a lot of light lambs on my hands. Everyone was struggling to come to terms with the new, post-subsidy economic realities, and the Meat Board was only offering around $1.20 per kilo, which meant that at roughly $10 per head, my lambs were only fetching half what the same animals had been worth the season before. The only bright spot was that wool was still king and, while meat was selling at a drastic discount, the clip was fetching around $3.70 per kilo net.

Efforts were underway to establish an export trade in live animals, and I heard a rumour that animals sent away under this scheme were fetching about $20 per head. I'll have a piece of that, thanks, I thought. I began asking how I could get a contract to supply the live trade. The following year, we landed a deal to send a few hundred lambs away on a live-sheep contract at the still very attractive price of $14 per head. It might have been at the same meeting in the packed Waipukurau Hall where Pete Wynn-Lewis declared that he was likely to have to walk off his farm that I heard the news that meat workers were planning to picket the wharves to prevent our sheep getting through. I let my heart rule my head and rose to my feet.

'Bugger that,' I said loudly. 'I've got some pretty wild dogs and a very harsh bullwhip, and I'm not scared to use either of them to get through the picket line.'

I received a huge ovation from the other farmers in the hall and, for a few days afterwards, older colleagues such as Dan von Dadelszen came up to me and told me how impressed they were to see a young fellow with a big heart in the game. Bay de Lautour reckoned I must have got my gift for impassioned oratory from the Bishop! Luckily, the strike dissolved soon afterward and we didn't get the chance to 'confront the enemy'. No doubt my passion and lack of wisdom and maturity would have seen me being arrested for breaching the peace at the least. So we got our live sheep away, even if we had to wait a long, long time to be paid.

While farmers were being asked to take a huge cut in their living standards (and in many cases, in the viability of their businesses) in order to save the economy, the cashflow cowboys in town were creaming it on the soaring stockmarket, rubbing our noses in it as they paraded about in their new BMWs and Porsches. Some farmers dabbled in shares. One day, Hugh Williams, a mate of mine, told me about a competition he and his wife had going.

'I've just bought a herd of 18-month steers to finish,' he told me. 'Adrienne's gone and bought some Robert Jones shares.'

When I next caught up with him, I asked how they were getting on.

Hugh scowled.

'The bloody Bob Jones shares are taking off,' he said. 'Adrienne's just made more money today than I'm going to get in the whole season fattening those cattle!'

It didn't seem right to either of us that you could use money to make money rather than have to work for it. But it turned out that Hugh and I were on the side of the angels in the long run. In the autumn of 1987, Hugh sent his steers through to the works and pocketed the meagre proceeds while Adrienne happily eyed her astronomical paper gains. If only she'd got out when the going was good. The stockmarket crash in September 1987 clipped her wings, as it did for a horde of other Kiwis who had been allowed to believe that the bull run would never end. The profit was an illusion.

Lesson to learn from this: the return *of* your capital is more important than the return *on* your capital!

Shortly before I took on the farm and while I was still working at Te Aratoi, I attended a free seminar that a local accountant's firm offered on keeping good business records. I found it hard to concentrate on all that talk of ledgers and journals because the session was being run by a very attractive young lady. Nearly a year later, when I was at the accountants' office to complete some paperwork on behalf of my sister, I saw her again. She was a bit older than me, and I was sure she knew a thing or two more than I did about how the world worked, but conscious that all work and no play makes Mark a dull boy, I plucked up my courage and asked her if she would like to go out some time.

She said yes!

We agreed that we would go to the restaurant at Vidal's, the local vineyard, and I couldn't wait.

When the day came, the dinner itself was going really well, with the conversation flowing as freely as the wine.

'Mr Warren?' So absorbed was I that it took me a moment or two to register the waiter hovering at my elbow.

'You're Mr Mark Warren?' he asked again.

'Yes,' I replied.

He grimaced.

'I'm sorry to have to tell you this, sir, but we have just been asked to pass on an urgent message to you. Your house has burned down.'

I stared at him, then turned to look, disbelieving, at my dinner date. She looked as shocked as I felt.

'We'd better have a raincheck on this, if that's OK with you,' I said to her, after what must have been half a minute where I sat in disbelieving silence.

'Of course! Yes.'

She reached across and held my hand.

'Are you OK?' she asked.

Blokes didn't cry in those days, especially not farmers, not even under the early days of Rogernomics. But I did feel a pretty deep gut-kick.

Why? I was wondering. And what the hell next?

CHAPTER 6

GRASPING THE NETTLE

After I had taken leave of dinner date and done a swift self-check to determine whether I had enjoyed too much of Vidal's hospitality to drive, I raced the Land Cruiser home in record time. When I got there, I discovered it wasn't my house that had burned down, after all. It was the bullock cottage, which was Hoke and Sandy's place. They were away at the time and therefore safe, but they lost everything in the fire. Our insurance company was great, but even so the payout wasn't enough to rebuild properly. Replacing the house would mean more debt.

Still, there was nothing to be done but grasp the nettle. I decided to embrace the principle expressed in the old Chinese proverb: 'Crisis and opportunity are equally weighted'. If we were rebuilding, we would make sure we built the kind of house we wanted on a better site. At this point, Dad stepped up. He might not have known much about farming, but he was very interested in building. We identified a couple of sites; when Dad eventually came up from Christchurch, he was even up at five in the morning a few mornings in a row to check where the sun was rising so that we could ensure the kitchen was bathed in sunshine at the dawn of a working day.

And then, when we began to prepare the site ourselves, using the tractor and the station's BTD 6 bulldozer, he declared that he was willing to learn how to use the machinery. I set him up with the new John Deere tractor so that he could rip into the hillside with an old, single-furrow plough, while I came along behind with the dozer levelling the building platform. It was the first time I'd ever worked alongside Dad (there wasn't much call for my skillset in his line of work, after all), and I enjoyed it. Mum pitched in, too, keeping us well supplied with her famous ginger crunch and lots of words of encouragement.

We could have done without the extra work and debt, but soon the builders at Waipukurau Construction were on the job prefabricating the new cottage and I could get back to the farm. The end result was well worth it, to the point where the fire episode seemed to be just a very well-disguised blessing.

When Hoke saw the new house coming down the road on the truck, he turned to me, his eyes damp with gratitude.

'Thanks, boss,' he said.

Our advisor from the local lime company has a good understanding of our type of country — for the most part, hard, yellow-grey Omakere clay that the locals tend to describe as 'friendly' (by which they mean it sticks to you come whatever, and doesn't let you go). It holds moisture well and carries a good load of phosphate, which is essential for pasture growth, but it has a low pH (that is, it is comparatively acidic) and won't yield up its nutrients unless the pH is raised. This is best done, as our advisor explained, by spreading lime, lots of lime. He also recommended we add a few key trace elements to the lime as well, notably sulphur. While sulphur is essential for promoting clover growth, which in turn is vital for producing the pasture quality necessary for fattening lambs, it can't be applied from the air by itself, because it is highly flammable and can ignite if it receives a shock. So it makes sense to add it to your lime.

By borrowing from the Napier branch of the BNZ — our local branch was Waipawa, but such was the scale of our debt that our account was administered from Napier — I was able to negotiate a pretty good price on the lime because, with times as tough as they were, most farmers had closed their chequebooks. I bought a thousand tons at $11 per ton. By putting the job out for tender, I was able to secure a very competitive price on the aerial application, too. The pilot who flew it on, Robin Langslow, took the job for $17 per ton. And the icing on the cake was getting a realistic rate on the cartage of the lime from the Hatuma quarry to the airstrip: an enterprising local farming family, the Butler brothers, were supplementing the income from their farm by using a few old Commer trucks to haul agricultural goods at a very low price. They hauled my lime for $8 per ton (these days, you'll pay anything upwards of $22), and while plenty of the existing transport companies gnashed their teeth and wailed about such cost-cutting, the reality was that they had got used to living off the fat of the land in the days of subsidised farming. There was no reason for farmers to be the only ones facing the new economic realities. Whenever I read about a transport company going under, I felt a twinge of guilt for my part in dragging down the market with my cut-price supply arrangements, but not for long. I reminded myself that everyone was simply adjusting to the brave, new post-subsidy world. I was just another victim of Rogernomics.

You can put fertiliser on pasture and the grass will grow (so long as it also has the right amount of water). But you need to be able to manage the movement of stock as well, protecting boggy or dry ground from the damage done by grazing (farmers call it 'mud management'). The art of farming is matching the feed requirements of animals at their different stages of growth to the state of the pasture, and vice versa. There are times when certain classes of livestock have to be fed as much quality feed as possible: you want lots of good feed there immediately after lambing or calving so milk production is maximised and animals can suckle their young, and once the young are weaned you need plenty of feed to give them a good start in life. At other times, such as in the winter, stock should be on a restricted diet, as too much feed can make your animals too fat or cause lambing or calving difficulties. That's why good fencing is vital to pastoral farming.

Within three years, with the repaired fencing and the application of fertiliser, we'd set Waipari up so that we could farm it a bit better. We'd done most of what we could to control the controllables, but other factors were out of our hands, such as rainfall and the markets.

It was plain to me that I needed to be smarter about the way we brought our stock to market. After all, it seemed pointless spending most of the year slaving away to produce the best animals you could, only to dump them into the market at the least advantageous time and at the lowest price on offer. Yet that was the way it had traditionally been done. The farmer had very little power over the price he got for his stock under the old business model. In the early 1980s, if you thought you had some good lambs, you'd ring up the stock agent or the fat stock buyer for your local freezing works and this man would drive out in his Holden Kingswood, often bringing a fresh loaf of bread or current newspapers. He would lean on the rails of your yards and cast his eye over the mob while discussing the latest rugby news, run your likely lambs up the race to feel the layer of fat around their tails, and mark them with a coloured raddle if they were 'killable' or (best of all) 'prime'. There was no set benchmark for either category: they often depended upon factors such as the weather, how much feed you had and therefore how much scope there was to grow more weight, how desperate the works were to fill their killing chains, how desperate you were for cash, or a combination of all of the above. Once he'd ranked your lambs, the agent would count out the killable ones, finish discussing the latest rugby news and tootle off back to town, calling the trucking company via his radio telephone when he got into range, or getting on the phone from his office.

The next day, a truck would turn up. You would load your stock, try to agree the tally with the truckie, remind him to give you a docket for it, and hopefully bid them goodbye. I say 'hopefully' because all too often you'd be seeing them again. If there was a drought looming and the draft had gone deep to get rid of as many killable lambs as possible, the meat workers' union would call a strike, knowing that with farmers desperate to process as many lambs as they could to lighten the feed demand as the dry set in and grass stopped growing, they had the whole supply chain over a barrel. Often some members of the union would see a chance to push for better working conditions or more pay. Sometimes this was a fair call, but often it was simply opportunistic and it would cause a big split in the community. (In more recent times, however, after the more militant unions' members realised that meat works could in fact just close down if the unions got too greedy, the situation has improved a lot.) If a strike was called, the lambs would be reloaded onto the trucks and sent home, often having had very little or nothing at all to eat and drink for a day or two, and having lost quite a bit of weight.

I realised that I badly needed to take more control of the supply chain beyond the farm gate, and the best place to start was with the stock agent.

In my third year, the agent who had been handling Waipari

stock for quite some time visited to look over that season's draft of weaner steers, which are ideal as replacements for stock that had been killed. However, I knew that the market for such animals was soft at the time, because a drought and a series of freezing-works strikes meant that farmers hadn't killed their older cattle.

'Don't worry about that,' the agent said. 'There are still a few buyers in the market, and I can get you an early slot in the sale where there's a firm buying order.'

I badly needed the proceeds from these beasts to pay the bills for the winter, so I decided to trust him. But on sale day at Waipukurau, I waited all morning for my animals to come up and they never did. Most of the afternoon passed, too, and when finally my lot came under the hammer, I looked around the thin crowd of disinterested-looking jokers sitting in the cattle rostrum, burping up reminders of the beer and pie the stock agents had shouted them at lunchtime, and felt a dose of rage and a sense of doom welling up in equal measures. The auctioneer yammered and cajoled, and while I could hardly understand his words, I could hear the numbers all too well. As he held his catalogue up ready to slap it in the palm of his other hand to signal the fall of the hammer, I piped up and told him there was no way I was selling at that price. That startled the cockies out of their boozy, post-lunch reverie! The auctioneer stared at me, open-mouthed. Then he shrugged. My cattle were passed in.

I felt my ears burning as I walked out of the cattle rostrum, with the crowd of much older and supposedly wiser cattle-men loudly expressing their astonishment at my recklessness. I found the truck driver and, fighting back tears of despair, asked him to take the mob home again.

On my way home, I dropped in at our local vet club to buy some drench to treat them for the upcoming winter. I was there standing at the counter trying to hold back the waterworks when Tony Thompson, a legendary cattle vet known to one and all around those parts as 'TT', wrapped a big hairy arm around my shoulder.

'Don't worry, mate. Come spring, you'll realise you just bought yourself some very cheap cattle.'

I wasn't so sure. The cattle were meant to reduce my overdraft, but here I was racking up more debt to buy the drench. And then there was the feed ...

It didn't bear thinking about.

A few days later, I arranged a meeting with the district manager of the stock firm, during which I expressed my dismay at the agent's performance. He listened, muttered a few platitudes, but seemed to forget about it the moment I walked out of his office. He might have paid more attention if he'd known I was walking out for good. And true to TT's word, I sold the cattle privately in spring for a price that was 30 per cent more, covering the cost of wintering them and then some.

The next time I was in the pub, the stock agent was there, and whereas he would once have shouted me from his tab when I went to pay for my beer, he turned the other way. I didn't mind that. In fact, I had already noticed how much business with stock agents was conducted over a few beers, and I made it a personal rule never to finalise a deal when the booze was flowing — unless I was more sober than the other party and I had the numbers written down. I had even taken to carrying a pocket calculator around, and I would occasionally duck out to the gents' to crunch a few numbers so that I knew I had my ducks in a row. It was a rule that I'm sure stood me in good stead in a great many deals.

Besides, the amount I saved in stock agents' commissions over time would have almost bought me my own brewery!

That episode set me thinking about the whole policy of just blindly following tradition and selling weaners in the autumn. Our winters out on the coast seemed a bit easier to get stock through than was the case further inland, so I decided that wintering cattle and selling them in the following spring was a better idea, even if it meant dropping cow numbers a bit. Previously the station had wintered 500 Angus cross-bred cows, but was tied into a system of feeding them from a self-feeder hay barn that had been built in the wrong place down in a valley. It got so muddy that the cattle got stuck trying to get there, and each summer we had to rebuild the road in so that we could drag hay trucks out there with the bulldozer. I decided that by dropping down to 300 cows and wintering all the progeny — about 250 calves — we would save hay cost and fetch a premium of about 25 per cent on the calves in the spring.

That was the whole reason we had the cattle on the property in the first place: when killed, their beef was only worth about

65 per cent of what lamb was worth. But cattle are workers on a sheep station. They keep the grass length down and the quality up in the summer so that the lambs can do their best on it. So one year, when we had a particularly good growth of grass in the spring, we kept the calves on to do their bit. By the end of the season, we found that in those conditions we could grow out good 18-month steers which were very much in demand from farmers who wanted to take them on to two-and-a-half-year-olds, and meanwhile the cull dry heifers were ready to kill on the local trade market by late May.

Now that I was thinking about how to maximise the return on our beef, it occurred to me that since bulls grow about 10 per cent faster than steers, the bull beef price is normally about 5–10 per cent higher. So I decided to try leaving 100 of our beef-bred calves entire (that is, not castrating them). I thought we could probably bring them up to 500 kilograms live weight by the time they were 18 to 20 months old in early May, whereupon they would kill out at around 250 kilograms carcass weight and be gone by the end of June, avoiding the need to carry them over a second winter.

Raising bulls is not without its problems, of course. Cattle tend to form smallish groups and establish a natural hierarchy. If groups mingle, the dominant animals from each group will feel the need to have a rumble to sort out the new pecking order. So our plan was to tag our 100 bulls with different-coloured tags, in groups of 25. The theory was that if there was a box-up, we could easily draft the mobs back into their original tag groups.

When I shared my plan with others, they were sceptical.

'Can't see too many black-and-whites out there on Waipari, boy,' they remarked. The bull beef market was mostly supplied from dairy herds, and the dominant breed was Friesians. My answer to this was that if Friesians suited our sort of country, everyone in Hawke's Bay would be running them. But they didn't. Cross-bred Angus, with their hybrid vigour, were hardier animals and a much better fit — never mind that they would typically yield about 3 per cent more in carcass weight, which translated to around $15 per head and a cost-free profit on the herd of $1500. Given this was fully 40 per cent of my annual drawings, it wasn't an insignificant factor!

The critics had a point, though; there was a slight flaw in my logic. The market was firmly of the mindset that dairy-bred Friesians were the only game in town, so if I didn't manage to bring my beef-bred bulls up to killable weight, there would be no market for them. Whereas an underweight dairy-bred bull could readily be sold as a store animal — that is, one that is bought to be fattened up to a killable weight — no one was looking for cross-bred Angus for this purpose. Interestingly, 25 years later, beef-bred bulls command a significant premium on the store market over the black-and-whites, precisely because they are hardier and produce a higher yield. Indeed, I have some very successful farming friends who used to finish steers but who now farm beef-bred bulls as a preference.

The other problem raised by the wise old heads that I consulted was that the young bulls would be forever vaulting the fences to get at the heifers, and I would be dealing with unplanned pregnancies all over the place. I thought this one

111

through, and decided that the problem would be solved if all of our 18-month heifers were in calf by 30 May. To this end, we would put six of our best (albeit narrow-shouldered, long-bodied) bull calves in with the yearling heifers on 10 October, for one cycle of 27 days. We would then scan them in February and anything that wasn't in calf would be sent down the road, either as a store local trade heifer or as one to keep fattening until the end of autumn. That way, all our replacement heifers were in calf as yearlings and we got an extra 45 calves a year. This in turn meant we could drop our cow numbers by about 10 per cent to around 270, all of which would do better on the available feed. We would be using only five instead of six stud bulls, so we could save on the associated costs (each stud animal costs around $5000 and lasts around four years). I was happy to forego the genetic gain that you achieve by bringing gene stock in from outside, and we would mitigate the expected loss of quality by earmarking the progeny of our home-bred bulls, so that the home-bred heifers could be culled and not used for breeding. But interestingly, after a few years we found that the home-bred heifers were just as big and well-grown as the supposedly better, stud-bred heifers. I don't know whether to chalk that one up to progress or luck, but some of our best cows have been home-bred. And in the end, all they had to do was to provide a calf that grew well and was hardy enough to survive the periodic droughts.

While livestock improvement is more art than science, one thing I soon learned was that the paramount consideration is how suitable your animals are to your conditions. I took things a step further and selected a South Devon bull to put over our mixed-age cows. (The theory being that cross-bred animals grow better on account of a quality called hybrid vigour: since the killing fees, levies and freight are all affixed to live weight, the heavier your killable animals are, the better your margin.) I found that I was getting bigger cattle all right, but that the South Devon-cross heifers were struggling to get in calf. Of a pregnancy-tested, in-calf mob of 65 heifers, only a quarter were Devon-cross. The rest were Angus types. Horses for courses: it appeared South Devons were simply too soft for our country.

My suspicions on this score were confirmed as I watched one of our neighbours buying in truckloads of hay to keep his big, exotic cattle alive in a drought, while our Angus types were managing without supplementary feed. My philosophy is that as soon as you start mixing diesel with grass (that is, spending money on the machinery and labour needed to feed out and perform other tasks associated with maintaining the condition of struggling animals), the economics of pursuing heavier animals just don't stack up.

When I took over Waipari Station in 1984, selling lambs for their meat was just a bonus. Wool was king, and it made up over half of our income. Shearing was a major event, with up to eight shearers and a gaggle of roustabouts, pressers and the like making gangs of up to 22 at times. The first morning's start at 5am was a real event, as it doubtless always had been on the property. The woolshed was built in 1882 and originally served the 22,000-acre Mangakuri Station from which Waipari was subdivided. There were two wings with 16 stands — rumour had it that originally, one was for Maori shearers and the other for Pakeha, and you can imagine the rivalry between the wings! The plaques on the walls attest to the many shearing records that were set in those days. The world shearing champion, Cam Ferguson, added his name to the wall recently.

We used only a single eight-stand wing. In those days, we budgeted for $4 net per kilogram of greasy wool: a good ewe would clip about 5 kilograms over two shearings. The relativities were pretty favourable, too. A kilogram of wool bought a gallon (4.5 litres) of petrol in 1985. A little over 10 years later, a kilo of wool would only buy a bit over a litre. Plenty of people have written about what has happened to the once-profitable wool industry, but suffice to say wool has now shrunk to about 15 per cent of our farm income.

In the wake of the removal of agricultural subsidies, I applied the same lateral thinking to the wool side of things as I had to the cattle. The fashion was to select the replacement ewe hoggets based on wool weight. Working on the theory that there could be a 2-kilogram variation in the genetic ability to produce wool, any extra produced over and above what you expected was pure profit, with no additional cost incurred apart from the extra woolpacks needed to ship it. But what we were in effect doing was selecting sheep with stronger (that is, coarser) wool, and we noticed that these were softer in constitution and were the quickest to fade in a feed pinch such as occurred in a drought year. Selecting for higher wool weight was therefore false economy: the strong-wooled animals suffered in our frequent droughts and struggled to produce twin lambs.

With the wool price falling to the point where returns were hardly covering the cost of shearing lambs, it was plain we needed to re-think what we were doing. My mates on South Island high-country farms didn't shear their Merino lambs in summer, so I decided to take a leaf from their book. I had worked out that when we shore our small, lean lambs before Christmas, we were being swept up into the panic and stress of getting the job done before Christmas and were exposed to anything that might prevent the shearing gang getting there on time. Simpler by far, I reckoned, to give summer shearing a miss.

This was greeted with derision from neighbours looking over the fence.

'The flies'll get 'em,' some warned.

'The buggers won't grow,' opined others.

The conventional wisdom was that lambs in wool over the summer months were far more susceptible to flystrike (where blowflies lay maggots in soiled fleece or in wounds, which proceed to eat the living flesh of the animal), and it was also well known that shearing lambs stimulates a growth spurt as their constitution seeks to establish a layer of insulating fat to compensate for the loss of their fleece.

But I had found some Ruakura research that called this idea into question. If you plotted the growth rates of shorn and unshorn lambs over a year, the study indicated, you'd find that the shorn lamb grew rapidly immediately after shearing and then slowed, whereas the unshorn lamb would grow steadily. The end result in either case was very similar.

And I thought there was good reason to believe unshorn lambs would be less rather than more susceptible to flystrike. We wouldn't have a whole lot of lambs running around with cuts and nicks from shearing in high summer, when the flies were at their worst. And it always seemed to be the case that we got a decent rainfall over the summer period, and this washed out much of what was left of the grease from the fleece. When we dipped the sheep to protect from flystrike (this was always the first big job back after summer), there was little or no grease for the chemical to bind to, weakening the effect. Despite dipping, we often found we had a flystrike problem in the lambs by late January. It was traditional at that point for farmers to blame the chemical company for producing dodgy dip!

One of the people I consulted on the road to dumping our summer shearing was my woolbroker (and mentor), Peter Nancarrow. Peter pointed out that the Chinese were paying a premium of about a dollar a kilogram for fine hogget wool (the technical criterion was that the average thickness of the individual strands of wool was under 33.4 microns in diameter). Because most New Zealand farmers aimed for heavier fleeces, stronger wool was preferred, and it so happened that by shearing young lambs, the fine 'tip' of the wool has been removed by the time they are hoggets. We soon realised that by shearing only once a year and later than was traditional, we retained the fine lambs' tip in the clip. When such a fleece was subjected to the air-flow test method used to determine micron, the presence of the fine-tip wool meant it averaged out to a much lower number, usually just squeaking beneath the magic number 33.4.

As a bonus, we also found that if we gave the lambs a crutch and good dip in early December, the dip chemical bound readily with the greasy wool and was far more effective. Indeed, I calculated we got three months' protection from a dip, and in so doing we drastically reduced the incidence of flystrike.

With the neighbours' sceptical laughter still ringing in my ears, we were silently banking an extra $4 per hogget from the increased wool value and saving the cost of a lamb shearing — about an extra $5 overall per head — to boot. By then we were wintering about 4000 hoggets, so the extra wool benefit paid about half the annual fertiliser bill.

Over time, of course, the loophole closed and the market niche evaporated, but in the meantime, looking over the fence on trips to town, it was interesting to note that there were a few other flocks of lambs being left in the wool over their first summer to become woolly hoggets. Imitation, as they say, is the sincerest form of flattery.

In the course of my research into the wool market, I noticed that our hogget wool was often described as 35F2D, which also happened to be the exact wool specification that was used as the benchmark in a wool futures contract. Futures provided a way of locking in a price, which could be a useful way of demonstrating a stable and predictable income to your bank manager. As a financial instrument, they had also become a bit of a plaything for the cashflow cowboys, the wheeling, dealing speculators who had proliferated in the mirrored-glass towers in the cities in the wake of Roger Douglas's deregulation of the financial sector. People with no connection whatsoever to the productive sector were making (and losing) fortunes dabbling in futures for everything from precious metals to coffee, pork bellies, beef and, of course, wool.

Because I was producing the actual, physical commodity involved in wool futures, I thought I'd have a go at making one of these deals work for me. I duly sussed out a futures broker to advise me and, after careful due diligence, I entered into a contract for a price that I thought was very worthwhile. I don't remember the precise details, but let's say it was for contracts to supply four standard 2500-kilogram lots at $6 per kilogram clean (which equates to about $4.70 net greasy — before the natural grease has been removed from the wool in a process called scouring).

A futures contract is, in essence, a promise made by a buyer to the seller to buy their goods at an agreed price at some time in the future. If the price falls below that stated, obviously the seller is the winner. If it rises, the seller is bound to the lower price and the buyer gets a bargain. It's all quite straightforward until speculators get involved, buying and selling the futures contracts themselves at prices that are based on their perceived value.

Over the five-month term of my contract, I watched with mounting satisfaction as the market price slipped closer to $5 per kilogram clean. But when it came time to close out the contract and realise my profit, my broker told me that the party on the other side of the agreement was seeking to limit their losses by negotiating a higher price — a no-no in the futures game. It seemed likely I wasn't dealing with a woollen mill, and when I did a bit of asking around, I learned that the other party was a tarseal cowboy with a big hat but no cattle (although he quite possibly had payments on a Range Rover to keep up). I found his details — the phone number placed him in central Auckland — and contacted him directly.

'Gidday,' I said. 'I've got some wool here that belongs to you. Where do you want me to deliver it? What's your address?'

The nervous-sounding bloke on the other end gave me a post office box number — a dead giveaway that I was dealing with a fancy-pants speculator rather than a wool buyer.

'What kind of forklift have you got?' I asked.

'Why will I need one?' he replied.

'Mate,' I said. 'I've got a truck and trailer here loaded up with about 10 tons of hogget wool that you contracted to buy and if you don't have a forklift to take it off tidily, I'll just have to put up the hoist and dump it in a pile outside your office.'

He didn't know that I knew that his office was in a mirrored-glass high-rise in Queen Street. He probably regarded that as need-to-know information only. He also didn't know that I wasn't actually behind the wheel of a fully-laden truck air-braking its way down the Bombays at that moment: that information was definitely need-to-know also! The contract was closed out within 20 minutes.

The futures broker was incredulous that I had managed single-handedly to manipulate the whole wool futures market (my price had been recorded as the ruling market price so subsequent deals went through at the same price).

'Ah well,' I said, '10 tons of loaded truck and trailer being backed up to block the door of a mirrored-glass building means a pissed-off cocky straight off the tussocks is always going to win out over a shiny-arsed cashflow cowboy who doesn't know the difference between a dag and a dagger, isn't he?'

'I suppose it does,' he replied, laughing.

The lesson here was that sometimes the 'little bit of force required' to finalise a deal is more efficient if you apply 10-odd tons at the first hit. I was very proud of the $10,000 profit I made on this deal, and put it towards buying a pure white (but very second-hand) Holden Commodore with 160,000 kilometres on the clock from our vet, Bert Middelberg. When I rolled into Waipukurau at the wheel of what was my first decent car, I began to feel at last that I might be getting on top of things.

CHAPTER 7

CAPTURING A RELUCTANT PROFIT

As mentioned, when I first took over the management of Waipari, I performed a SWOT analysis on our operations. Our strength was the fact that we could produce hardy sheep that did well in harsh conditions. While our summers were normally dry and lacked the vigorous clover growth needed to fatten lambs quickly, our comparatively warmer winters meant that if we could carry lambs over the summer, we could then optimise them on the autumn flush and deliver them into the emerging winter market, when supply was dropping and prices were rising. The weakness was that we somehow had to get the lambs through the summer, and while we didn't need them to grow quickly, we had to feed them more than just maintenance. The only way I could see to do this was to reduce ewe numbers. Talking this through with a number of advisers, they indicated that I would probably find that I would get at least as many lambs from 4000 ewes as I had been from 6000, because there would be more and better pasture to go around.

So it proved. But it was plain to me from sixth-form economics that there was no point in doing what we'd always done — producing small-framed, 'second-choice' grade store lambs at precisely the same time as everyone else's stock reached the market. And in 1986 when the lamb schedule was announced, the prices on offer were very low. There seemed to be no point in raising lambs for meat at all. Since wool remained relatively buoyant, I decided we would have a go at South Island high-country-style wool-only production. We would keep an all-wool mob of 2000 wether (castrated male) hoggets, the theory being that wethers are under less biological stress than ewes and therefore produce a superior fleece.

This was going pretty well. We had a good autumn, and the mob was coming along nicely. But in the early spring I got a phone call from Roy Fraser, one of our farm advisers, who was in contact with the Heavy Lambs Trust. The Trust had arisen from the initiative of sheep farmers John and Sara Atkins, who had sniffed out a market niche in the shape of a considerable demand for heavy lambs in the United States. In late 1986 they left their farm and moved to San Francisco, where they established an operation selling lambs 'direct-to-butcher'. To supply them, John Atkins and Phil Guscott assembled a group, which was registered under the name Heavy Lambs Trust, which was made up of Wairarapa farmers who had found themselves in much the same situation as we had.

Knowing we were making heavy wethers of our mob over winter, Roy Fraser phoned to make an offer on behalf of the Trust. Under their pricing, the hoggets were worth more when they had been shorn than they would have been if we carried them through to autumn. Well, that didn't require much deliberation on my part!

I was intrigued, and made an effort to find out where these animals were going. I did a bit of due diligence and learned that the Trust had identified a demand for lambs of between 18 and 24 kilograms, with a GR (that is, the measurement of the layer of fat around the seventh rib) of between 3 and 12 millimetres. I also learned that there was normally a premium of 15–20 per cent payable if you could supply them in the early October period. The only potential threat I could identify was that we might produce larger lambs than the market required — a lucky problem to have, as quality lambs were in short supply on the local market at that time of year, so the schedule price was usually quite favourable.

I contracted the majority of our lambs that season to the Trust, which had them killed at the facility belonging to Craig and Penny Hickson of Progressive Meats in Hastings, and then exported to the United States. In 1989, when the Heavy Lambs Trust was re-incorporated as the Lean Meats Company, we bought shares in order to secure preferential killing space. The plan was to hold all our lambs over the winter and supply them to Lean Meats via Progressive Meats' plant at about 500 per week, starting in the last week of September and finishing just before the Hawke's Bay A&P Show, which just happened to be at exactly the same time as we were docking, ewe hogget shearing and trying to get some spring crops in the ground. Those were big days and high-stress times, but the reward was well worth it.

As my neighbour James Aitken constantly reminds me, there are three ways to make a farmer change the way they do things. The first way is to pay them more; the other two don't matter. Farming has changed a great deal since I took on Waipari. Mostly this has been driven by the sheer necessity of finding better and more lucrative ways of doing things. One of the consequences has been the decline of the Stock and Station Agency. In the old days, bulk farming inputs were delivered by coastal trading ships, which also picked up any wool. The stock firm held the ledger in Napier and managed the finances, paying the accounts from wool and stock proceeds. They took a commission on most stock sales of about 5 per cent, with a trusted stock agent acting as a farmer's eyes and ears as to where best to sell stock.

Over time, this system evolved to the point where you could buy most things — from groceries to tractor parts — from your local stock agent, and there were other informal benefits, too. If you were in good standing with your stock agency, you could front up and ask for an advance from the manager's petty-cash box to spend at the pub that night if you were a bit short. I once managed to book up a bottle of rum at the family firm Williams and Kettle, but those supervising the farm accounts delivered a stiff rebuke, judging it not to have been an essential farm input. What would they know?

After taking over at Waipari, I soon realised that paying a 5 per cent commission to an agent was a reducible cost. After all, most of the time, as already described, all they did was drive out, look at a mob, value it and suggest it be sent to another farmer, while I had done most of the work.

I also had good reason to believe we were missing out on some of the value of our stock by leaving agents in charge of assessing it. Early on, I decided to make my own assessment, and I hefted a lamb and climbed onto the bathroom scales so that I could calculate the difference between Mark + lamb and Mark alone in order to determine the lamb's weight. I did this for a few animals so that I could arrive at a rough-and-ready average live weight of the mob. My figures seemed to show that our lambs were being underestimated by the agency.

So I bit the bullet and shelled out for an expensive set of electronic scales — one of the first in the district. With an accurate grasp of the weight of our stock, I could use the market price per kilogram for livestock published in the newspaper to value them.

The first time I used the scales to full effect, they justified their purchase price. We had a mob of 500 cull ewe hoggets in the wool to sell at 32 kilograms a head, live weight. I fielded an enquiry from a local private stock agent who was buying on behalf of one of the big meat companies. This was a very significant sale for us, and I was determined to make sure we got full reward. To get an exact idea of the value of the hoggets, I had selected four, weighed each of them, shorn them, weighed the fleeces and then killed them for special house muttons, carefully weighing the dressed carcasses to give me an accurate idea of the average yield (live weight to carcass weight).

By doing that I had a very good idea of what they would be worth if I killed them on schedule (for the advertised market price). The agent arrived, promising a 'good premium' over schedule. He spent a while looking them over, then turned to me.

'What do you want for them?' he asked.

'Eighteen bucks a head, I reckon,' I replied.

'Piss off!' he spluttered. 'They're not worth more than $14!'

I explained how I had arrived at my estimate of their value.

'You've got to be bloody joking,' he said. 'I will personally shear the buggers myself and I guarantee you will not get more than $14 a head for those animals. I'm a busy man. I'm not here to be pissed about by young bucks like you!'

I shrugged and told him my price stood. He got even ruder, and I asked him to leave the property and never set foot on it again. I killed the lambs on schedule and got well over $17 per head for them.

As a result of farmers tightening up their practices and slashing costs, many of the small service companies in the rural towns struck hard times and closed their doors — and not just the small companies, either. When I came to the Hawke's Bay district in 1983, there were five stock and station agencies in our local town, Waipukurau. Now there's only one. The old family firm, Williams and Kettle, formerly the biggest in town, shut up shop, and the premises it formerly occupied have now been taken over by the Waipukurau branch of New Zealand Work and Income. Now that's a sign of the times, if ever there was one.

CHAPTER 8

RISK AND REWARD

Farming, like any other business, is all about risk and reward. You're constantly required to make decisions that affect your business, and the security of all you have invested in it. When, occasionally, I go to the local races, and someone asks me how my gambling is going, I always reply that I seldom bet more than a dollar on a horse (mainly due to that fact that the odds are always stacked in favour of the IRD and the TAB). I make hundred-thousand-dollar bets in my farming business every day and don't want to do it for fun, too! The reason I go to the races at all is to catch up with other businesspeople socially, and hopefully pick up a few pearls of wisdom.

I would be being less than truthful if I were to deny that I have made many mistakes, some minor, some major, in over 30 years of being in business. It's never a healthy thing to dwell on one's mistakes, but if you learn from them and put processes in place so that they're not repeated, you can find the silver lining in most stuff-ups. It's been well said that 'the best lessons are the expensive ones because you learn them quickest'.

Every now and then, you have to stick your neck out. In my book, doing nothing is a bigger crime than having a go and being knocked back from time to time — so long as the mistakes are learned from and not repeated. Many is the time that I have attempted to outguess a market trend and have taken a position that has proved to be less than ideal. It's easy to respond to a market signal, such as where prices are rising, by holding onto a product (wool, say) in the hope it will go even higher, only to have the law of supply and demand kick in with market resistance causing a drop in demand and price — a clear signal to others to dump more product on the market, which has the effect of further reducing prices. This is what is behind the so-called 'boom and bust' cycle you so often see in commodity prices. But with the backing of quality market information and armed with a basic understanding of economic cause and effect, I haven't been caught out too many times. I have learned the hard way to sell when the market is offering what I regard as a good and fair price, and to resist the temptation to try to sell at the peak.

Probably my worst judgement calls have occurred when I had too many jobs on at once and wasn't giving clear and careful consideration to the quality of my decision-making. At times like these, I've also been too trusting of advice offered by someone keen to sell their wares, for example, only to find I've been landed with some crap. Sometimes the best way to work out whether a product is any good is to ask someone selling an alternative what they think of the product you're considering. If they admit that the opposition product is good, you've learned two things: you've got an honest appraisal of the product, and you've found a good, honest adviser.

From time to time, I've been seduced into putting hard-won cash into investments that promised a much higher yield than farming, whether it was shares or direct investment in other people's businesses. Almost without exception, the investment has earned less than the opportunity cost of the capital — that is, the value I'd have got from my money if I'd simply paid off debt. In one case, not only did I not get a return on my capital, I didn't get the capital returned at all! So I have learned two things. People's personal assurances and guarantees are often not worth the paper they are written on, and it's best to stick to your knitting. There's an old saying that you should beware of wise men bearing gifts, because if they're as wise as they're cracked up to be they won't be giving anything away.

<hr />

While we were having some breakthroughs in the economics, like all businesses farming comes with its own peculiar set of risks. With long, arduous hours spent outdoors, often on your own with heavy machinery and cantankerous animals, some of the risks are more peculiar than others. Most farmers have some pretty hair-raising work stories. Sometimes I think I have more than my share. As a recreational user of adrenaline (that's what 4WD rallying is all about), I've also been in many situations where it has served its intended natural purpose: namely, keeping me alive!

Once in the early days, while I was driving a mob of two-tooth ewes home, I approached a point in the laneway at which we would have to cross a deep gully by passing over a narrow culvert. There was a bit of scrub just to one side of it and, as I came around the ridge, I saw many of the ewes were attempting to cross through the scrub, not over the culvert. They were emerging from the scrub and falling into a deep hole below. Sheep being sheep, those coming behind were simply following the leaders, traipsing through the scrub and then falling on top of the animals already floundering in the pit.

The technical, farming term for this is a 'smother', because that's what happens. The sheer weight of the bodies above them smothers the sheep trapped below.

The standard stockman's procedure is to tie up your dogs, so they don't push more sheep into the problem, and then do what you can to cut your losses. Two-tooth ewes were probably the most valuable of any sheep on the place, and they were in lamb. I was buggered if I was just going to stand by and watch them all expire in a struggling, groaning mass. I did my best to divert other sheep lining up the deathly hole, and at the same time I tried to drag as many suffocating sheep from the pile as I could. As each weighed 60-odd kilograms, this was more easily said than done.

The silly sheep just kept following their woolly sisters and jumping in on top of me. After I had ejected around 30 ewes, my strength was about gone. There was a sudden new influx of woolly lemmings, and all went dark — I was buried under hot, woolly flanks and flailing hooves.

Like most people, I've always had a morbid fear of suffocation: perhaps I had been stuck at the bottom of too many rucks in my schoolboy rugby days. This was a nightmare coming true. Despite spending my childhood in a vicarage and my school days under religious instruction, I had seldom resorted to prayer. This was one time when I did. And my prayers were answered. The last of the mob tumbled into the pit, and with what was left of my strength, I was able to wriggle out of the ovine scrum.

After 10 minutes spent getting my breath back, I found I could spare a few puffs to administer mouth-to-mouth to a few very groggy ewes. (Actually mouth-to-nose: you clamp the animal's mouth shut and seal your lips over their nostrils and blow. If you think it sounds unpleasant, you have no idea, but in the end, I'm a farmer and all I had to do was think of the best result.) When everything was back on more or less wobbly legs (mine were no different), I was able to gently get the mob moving to the next gateway in the laneway so the sick ones could be put into a paddock to recuperate. I didn't lose a single animal.

I still get a cold chill when I remember the feeling of being buried under all those sheep, though. All it would have taken was for another three or four to blunder in and I'd have been a goner.

Another incident in the early days came when I was mustering the coastal country. For reasons which, out of consideration of trying to maintain family harmony, will remain obscure, there was a stray stallion roaming around and it took an immediate and obvious liking to the mare I was riding. He made it plain that he meant to have her there and then, and he wasn't at all fazed by that fact that I was in the saddle.

I had seen a number of blue movies in my student days at Lincoln that had convinced me that human beings have no place in an equine threesome. I shooed him and whacked him on the nose with my rakau (a makeshift riding crop, in this case comprising a length of alkathene pipe). So far as you could tell, all this served to do was pique his interest. He reared up, his forelegs clamping the mare's quarters and, breathing heavily down my neck, he prepared to go to work.

Then Dylan sprang to the rescue.

Like most gentlemen, who are trained to finish fights rather than start them, he dashed forward and sprang about 5 feet into the air in order to sink his teeth into the fleshiest part of the stallion's shoulder. Then, with a firm grip, he twisted his head, which certainly spoiled the mood for the stallion. The horse writhed to one side to try to ease the discomfort. This unbalanced him, and he slipped off the mare. Dylan hung on until the animal staggered and fell, whereupon he scooted out from under several hundred kilograms of thwarted lover and we all galloped for the nearest gate. With the gate between us and the stallion, there was time to see the funny side. I wished there had been a camera there — I could just hear

David Attenborough's commentary — but I wasn't about to try to re-enact it.

※※※※※※※※※

Speaking of animal programmes, I later learned that making TV can be very frustrating for dogs! One time when *Country Calendar* visited, I was using my team made up of Tom (the heading dog) and huntaways Diva, Meg and Bessie (the very biddable huntaway of my first wife, Chris) to drive a mob of hoggets up to a gateway. The crew had only one camera operator, so despite the fact that the dogs had executed a perfect mustering manoeuvre, once the mob was through the gate the director Julian O'Brian called 'cut', and said, 'Could we then get another shot from the front, with the sheep coming through the gate, to balance the shot of the side view?' I told Tom to head the mob and bring them back through the gate and to re-enact the whole scene again. The dogs were confused, as I had praised them wholeheartedly for doing such a perfect job and making me look good on film, when in reality the opposite would have more often have been the case! They obliged, but you could see them thinking, Why do we have to do this again when we did it right the first time? Tom was also a bit peeved at having his great 'head-and-pull' wasted as we reshot the scene. But he got with the programme and did as he was told.

'Such a great shot!' called Julian, as he ordered the obligatory 'cut'. Then he asked for another shot, from a long-distance perspective! I couldn't believe it. This time I pushed my luck and

My family with the 'Queen Mum' in Timaru, 1966. I'm the blond one in the middle.

Me in Waihi, aged about nine, 1969

My first go-kart, built out of wood and hand-mower mechanicals, 1970

A ski group, packed into 'Truckie', a 1956 Vanguard ute, heading off for a ski week in Tekapo, 1976

Me and my 'Ski Dub', 1977

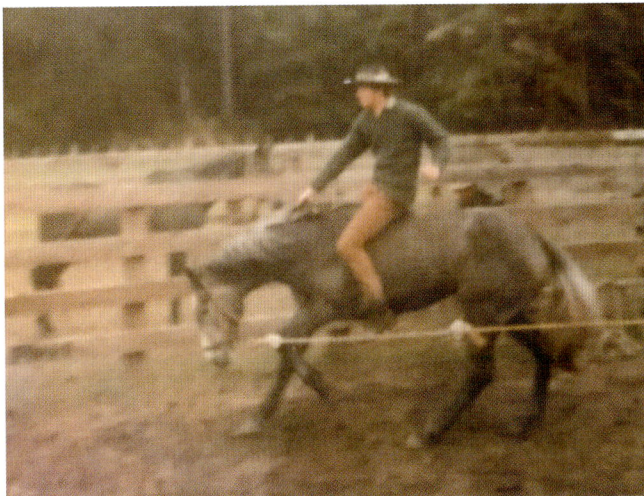

Learning how to break in
a horse with Linny Morris,
Okuku Pass Station, 1978

On the Mount Cheeseman
skifield dozer, 1980

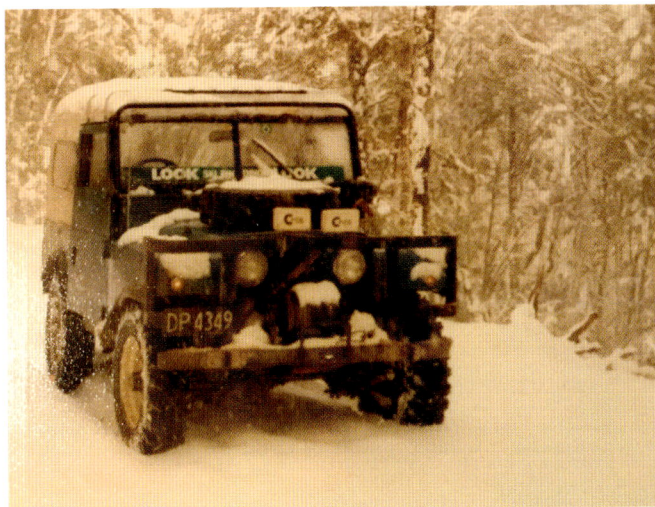

My first Land Rover,
Jumbuk, on the Mount
Cheeseman road in 1980, on
my way to rescue a stuck car

In Jumbuk, crossing the Waimakariri River, 1979

I'd pushed my luck with the Jumbuk's deep-water abilities this time! Waimakariri River, 1980.

Arriving home in Napier on 29 June 1984 after my world trip. Dylan was waiting in the airport car park to greet me.

Dylan and me asleep on the shingle, December 1984

About to head off on my first tour of Waipari after taking over management, 1 July 1984. Tom, the heading dog, was quick to claim the seat with the best view.

Me with Dylan and my mother in the Kangaroo, a much-modified Land Cruiser, 1987

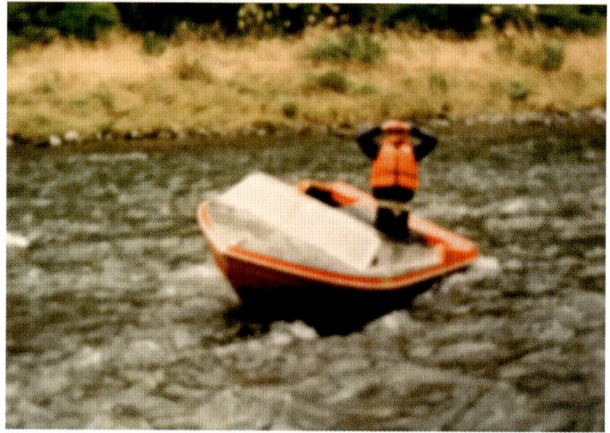

High and dry in *Cardiac Arrest*, Rangitikei River, 1991

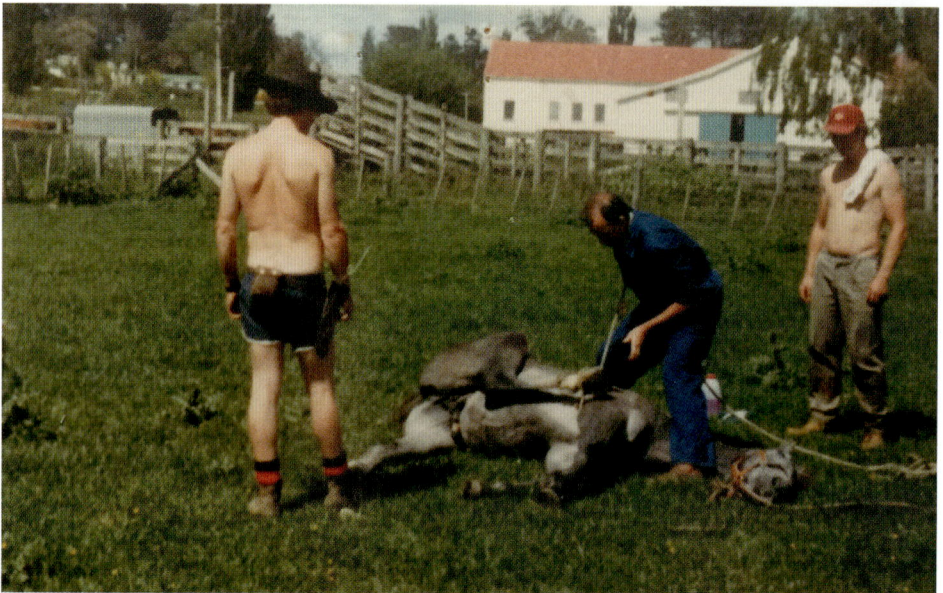

Watching legendary vet Richard Lee tying up a young stallion, with Tony Mackie looking on, 1984

A very clean Jumbuk diving into the Waipara River at the Quadrive 4WD Club rally, 1980

Winning the 1988 V8 production-modified NZ 4WD rally series in the Manawatu Overlanders Rally. The overall winner tried to claim I hadn't finished due to the fact that my front hubs didn't pass through the blue pegs as the rules stated. It's true: they passed over the top of them by some distance!

The Land Cruiser leaping into the air. To achieve your goals, aim for the sky — but keep your driving wheels firmly grounded.

My Nissan Patrol setting up the track for a Freezedrive course at the Cardrona vehicle testing ground, Snow Farm, 2003

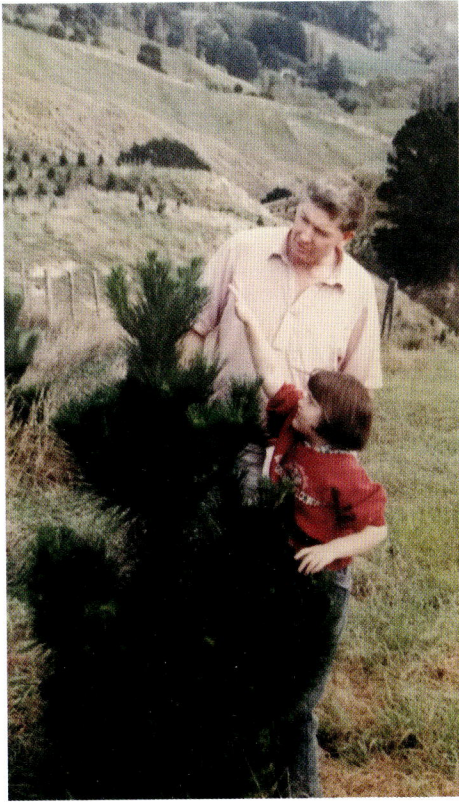

Helping my daughter, Emma, choose a
Christmas tree, December 1995

The Warren boys and Marley, 2010

The trusty 1979 Land Cruiser on the main Waipari access track. Just another muddy morning!

Coming along the Mangakuri road on a rainy afternoon in 2008. We need a 4WD just to get home sometimes!

Lucky we had the Patrol!

Hold your breath, it's getting a bit deeper!

Discing in forage barley to finish winter lambs in Henry's Paddock, 2016

A prime load of Waipari A-grade logs being taken off to be made into furniture somewhere in the world, January 2015

Just getting to work can be a 4WD challenge in itself, but it's better than being stuck in a traffic jam! *(photo by Lee Warren Fotoshoot)*

Some days, gravity wins outright. Another muddy morning, driving in fenceposts while the ground was soft enough, July 2014.

Running a police Freezedrive course at the Remarkables skifield car park, June 2009

This is Skidsprint, my new motorsport. Here I'm pushing the boundaries of gravity in my Maverick, at the opening of our motorcourse track, 'LeMarc', November 2017. *(photo by Jeff Wells)*

Henry learning how to cross a muddy ford, in his much-modified Suzuki SJ413, July 2010

And learning, the hard way, how to dig it out if he gets it wrong!

Henry researching the limits of the Suzuki's water-fording ability — and wishing he had fitted a tow rope before he drove in, 2012

Jack racing home for lunch, 2010. The boys are allowed to push their luck in a vehicle to learn to drive properly, as long as they have their seatbelts on.

Henry and Jack having just found the limit of the Suzuki's roll point, 2010. Experience is the thing you get a split second after you needed it.

Jack, William and Henry paying their respects at our own private dawn service at the man cave, Anzac Day 2010

The track to my private clifftop man cave. I don't have an address, just GPS co-ordinates.

My retirement project, a totally original 1953 Series 1 Land Rover called Jock, 2015 *(photo by Lee Warren Fotoshoot)*

A sunny afternoon at the man cave, 2015 *(photo by Lee Warren Fotoshoot)*

Dawn seen from the man cave, 2015. This photo was used in the New York World Trade Centre launch of our Just Shorn wool brand. *(photo by Lee Warren Fotoshoot)*

Julie and me at the man cave *(photo by Lee Warren Fotoshoot)*

the dogs' patience too hard. Tom did a very half-hearted head, leaving a few behind, and Bessie ran off to snatch a congratulatory pat from Chris, who was observing from a safe distance with the rest of the camera crew, while Diva went off to find a cow pat to roll in, leaving only Meg to re-enact a somewhat untidy muster. And to add insult to injury, the film editor cut the shot. I was hoping to show it to the dogs afterwards to explain their mistakes to them, but no luck!

In a later scene, I had Tom with me as Sam Kingston and I drove a mob to the woolshed. I spent the whole time calling Tom in close to me, as he had an ongoing feud with one of Sam's dogs. While Tom was a meek and mild dog around humans, around other dogs he had developed a set of killer jaws and an attitude to match. His mouth had to be prised open with a stick or his head held under water before he would let go of another dog. The vet bills had been mounting, and normally he had to wear a muzzle if he was around other dogs. Annoyingly, on that occasion it had been lost, and so, in the now slightly infamous film shot, the realistic sound clip recorded through my mic was later revealed to be full of expletive-laden threats as I tried to prevent a mid-scene dog fight. Hence, the truly realistic soundtrack had to be edited out and replaced with the director's voice-over. The day before, while attempting to save another innocent dog from Tom's attack, I had held a terrified, quivering dog above my shoulders to keep him out of harm's way, only to have Tom give me a very solid nip in the ribs as he tried to add another notch to his kennel wall. Luckily, my good thick woolly Swanndri protected me from the worst of the bite, but I still had a big bruise and teeth marks to show for it!

On another occasion we had been mustering heifers out of the Ware Paddock, a 250-acre rough back-block. The job had taken me most of a winter's afternoon. It was just after dusk, and with the best of the light well gone I got the mob of about 55 heifers up to the corner gate and into the laneway to be collected the next day. I went in ahead to open the gate, and called Tom to walk up on the mob and bring them towards me. At one point I heard a bit of a kerfuffle, with the dogs barking excitedly, a few strained moos and bit of a dog yelp. Continued calls to Tom to walk up on the mob weren't met with any action. In my frustration, and as darkness set in, I gave a more determined command for Tom to hurry up. I couldn't see him. Normally the big white blaze on his chest stood out well in all but the darkest night. No sign. I turned the bike around to use the headlight and immediately spotted him, walking weakly with his white chest blaze a dark bloody red. He was trying his best, but he had lost a lot of blood and was very weak. I told him to sit, took off my Swanndri and wrapped him in it to reduce the shock and cold, and with Diva, a very young, inexperienced huntaway, and lots of motorbike horn, I managed get the mob up through the gate before I returned to nurse Tom. I reassured him that he would be OK, and just to hang on and not die. I scooped him into my arms, balanced his listless body across my knees and rode as fast as I could along the 6 kilometres of track back to the homestead, leaving Diva to follow as fast as she could. Reaching the homestead, I carried Tom inside to keep him warm beside the fire while Chris gave him first aid to stop the bleeding and I rang the vet.

Picking up the party-line phone, I heard the relaxed female chat of Sandy catching up on news with a friend. With no time for pleasantries, I said: 'Quick, Tom is dying, please get off the phone so I can ring the vet!'

Luckily I was able to get hold of Ian Walker, one of our great local vets, and he agreed to meet me at his practice ASAP. Wrapping Tom in another sheet and blanket, I laid his bleeding and muddy body on the pristine woolly seat cover of my new Commodore.

A very rapid 25-minute trip to Waipukurau — with me urging Tom to hang on, stroking him with one hand while I drove with the other, and assuring him that help was on its way — had me meeting Ian, who was ready to go with all the emergency support.

Ian's expert care, drips and a few days in the vet club had Tom back on his paws again and pleased to be home, albeit with a brisket full of stitches. The day after the incident I traced the trails of blood and found that Tom had run into a sharpened, broken-off manuka stump, probably as he was focusing on heading off an escaping heifer.

Later on when that block was subdivided, Tom was honoured in perpetuity by having that paddock named after him.

In the national museum, Te Papa, is a Swanndri belonging to another Hawke's Bay farmer, who I understand used his iconic woollen bush shirt to save his favourite dog in a similar situation.

To keep up the numbers, and perhaps stop a more serious young dog triallist from suffering the embarrassment of coming last, I used to enter occasionally at the local trials. Out of respect for Tom's enthusiastic ego, I entered the 'long head and yard' at the Omakere district trials.

The general plan in this event is that man and dog stand in a marked circle about 20 metres across and attempt to muster a trio of three sheep, released quietly from a yard some 200 metres away up on a hill-top. The sheep are from a flock of about 100, held in a well-camouflaged yard around the side of the hill, well away from the dog's sight.

A couple of experienced shepherds select the three sheep and coax them around the hill, helped by well-positioned temporary fence netting, so that the waiting heading dog can see a group of three sheep and make it its focus to round them up. The dog's job is to bring them gently and smoothly, in as straight a line as possible, towards his master in the ring. When the dog has brought the sheep into the ring and the sheep stand still in it, the combination of man and dog are judged to have completed the job, with final points awarded for various rules of perfection, such as having the sheep settled and not trying to bolt away.

Tom stood beside me in the ring, anxious to get the job underway. His eyes scanned the surrounding hills for places where sheep could have been hiding from his strong-eyed gaze. With the three sheep waiting patiently on the hill-top, the judge called 'time' from his shady tin shack.

At the utterance of the magic command 'Way, Tom', which was meant to mean that he cast widely out to the right, he sprang into determined action. Having already scanned the hill and worked out himself that the right-hand side of the hill was clear of sheep, and aware as always of the boss's mantra 'Time is money', Tom ran straight to the waiting three trial sheep, mustered them around the hill to their sisters, jumped the scrim fence of the holding pen and firmly encouraged the waiting 100 or so to join the selected three, causing a flattened fence in the process.

There is a certain unchallengeable etiquette you have to display when standing on the dog-trial course and good manners (and certainly no swearing) is near the top of the list! There was no question that Tom had performed an incredibly efficient mustering manoeuvre, but he had by-passed the spirit of the game and not left any sheep behind for the next team! His draw towards the ring was faultless for its straight line; however, the settling aspect had been overlooked and, as the mob charged at the 20-metre wide ring, there wasn't enough room for all the sheep plus Tom and I to fit at the same time. Trying hard to suppress my embarrassment at the damage we had caused, I called 'Wellago' (come away and let go) to Tom, thanked the judge for his time, apologised for the mess we had created, called Tom in for a reassuring pat, then went and hid behind the ute for half an hour while the officials relocated the wayward sheep. It was a while before I felt I could safely sneak into the bar through the back door without attracting too much ridicule!

On a later occasion, at the Elsthorpe trial, Tom's reputation seemed to precede him, and as the three sheep were shepherded out to await his mustering skills, I noticed that they would not settle. However, at the magic command, Tom shot off in his usual enthusiastic manner, only to discover that as soon as he found the sheep and walked up on them to move them, the sheep fell over — in unison! It turned out that, as a joke at my expense, the three had had their legs tied to each other with bailing twine and were effectively immobilised. Tom was bemused, alternating between looking at the belligerent, unmoving sheep, and looking to me to provide intervention.

Tom wandered back, dejected that his infamous technique had failed. I got the hint, accepted the jibe and, as I called Tom off, retired gracefully to the bar. We gave up that sport after that, with Tom and I agreeing in silent communication that perhaps only doggy people can understand.

It would be easy to write a whole book about dog stories, but like favourite children stories they are normally of interest mainly to the parents. So, apart from a couple, I will leave the serious doggy stories to the committed dog triallists!

When we switched from raising steers to raising bulls for beef, we increased the number of large, potentially dangerous animals on the farm — although in reality a cow with calf at foot can be a far more dangerous beast than a bull, if you should get between them. A cow will charge repeatedly at anything she perceives to be a threat to her calf, be it dog, man or even sheep. In my 35-odd years farming, I have been charged by about three times as many cows as bulls. On one occasion, I didn't quite get away, and I carry a significant knee injury as a consequence. I climbed the rails of the cattle yard to escape the angry beast, but she still managed to push my knee sideways through the rails. Even now, despite having had many operations on it over the years and, more recently, a full knee replacement, I walk with a limp. Every now and then, we get a cow who will charge a person for little or no reason. Such animals get a one-way trip to McDonald's.

All the same, no farming yarn would be complete without a raging bull story or two. On one occasion, while I was trying to get three 18-month-old bulls through a gateway, I pressured them too hard and ignored the hierarchy of the trio. The lowest in the pecking order was hanging back in order to avoid being in the gateway at the same time as the boss bull, for fear of getting rammed into a post for his troubles. To speed things up, I pushed the front wheel of my motorbike up to his rear end to give him a hurry-along. That had the opposite effect to that intended, and he swung around, lowered his head and came for me.

I sprang off the bike and hid behind a very large, conveniently placed thistle. While this would have offered no protective value whatsoever if he'd switched his focus onto me, it served to remove me from the picture and leave my hapless motorbike at his mercy. He set to trying to trample it into the ground, ramming it repeatedly with his head, until his lips came into contact with the hot exhaust pipe. With a pained bellow, he abandoned his metal foe and trotted off through the gate to join his mates. Mission accomplished — albeit at the cost of a lavishly bent set of handlebars.

Another time, I was droving a mob of about 15 bulls along the road. As it was a very hot day, they were a bit inclined to drag their toes. A group of three had decided to take the path along the fenceline between the long, rank, overgrown grass and the eight-wire fence, with me ambling along behind them. All was well until we came to a culvert crossing with a bit of a ditch to jump. I gave them a minute or two to look at it and work out how they could cross it, but they appeared to have decided it wasn't jumpable.

I came up behind them and grabbed the tail of the smallest one — a mere 500-odd kilos — planning to twist the tail to encourage him to jump over. The other two wheeled and ran back past me. I wasn't going to give in that easily, so I gave the tail of the last bull an extra twist to encourage him to take some leadership.

He objected and turned on me.

If an animal is trying to charge you, the golden rule is to grab its head and hang on, covering it so that it can't back off and get a charge going at you. It also gives you a chance to use your arms to cushion the impact. That's the theory. In practice, I didn't have time to do that. As he wheeled, I backed away, but the effect was to give him a 5-metre run-up. Finding he didn't have any space to escape with dignity, he decided I should be on the other side of the fence. He dropped his head, rammed me into the fence and began working me up it with short, vicious flicks of his neck. That would have been bad enough had the second-to-top strand not been barbed wire. The barbs snagged me as the bull smeared me ever higher. Then he dropped his head and gave me an extra vigorous flick that send me flying into the paddock beyond the fence, where I landed flat on my back a few metres away.

The upside of this was that the bull immediately lost interest in me, and trotted off happily to join his mates, leaving me lying very still and sore on the hard, rough ground of the neighbour's paddock.

My four dogs, who had all been warned by the bulls not to push their luck in the past and who had therefore been prudently hanging back, came over for a sympathetic sniff and to see why I was suddenly very quiet.

Gradually, flexing my various toes and fingers, I decided that my injuries were more external than internal, and I slowly rolled over onto my side to attempt to get to my feet.

Everything hurt.

Just then, I heard a vehicle drive slowly past. They couldn't see me because I was lying behind the long grass adjacent to the fenceline, and I couldn't get up in order to attract their attention. I belly-crawled to the fence and used the wires to pull myself upright. The vehicle was gone up the road, but I was pleased — and moderately surprised — to find that nothing major seemed to be broken or sticking out.

What to do next? If I could make it to my bike, I could ride home, but what good would that do me? There was no one there. Would I ring 111? I didn't seem badly enough injured to warrant that. And simply riding off would leave the mob of bulls loose on the road and make the dogs run the 3 kilometres home without finishing the job.

So what, then? Tie them all up? I cringed. Even the thought of trying to get them all sufficiently under control to secure them hurt.

Then the dogs, who had been confused initially, seemed to grasp the situation and took control. They gathered the bulls up and began pushing them towards the yards, as if to say: 'Why didn't you let us do it our way in the first place, boss?' I decided to finish the job, although I could feel blood running down my back and legs. I left a trail of blood as I rode slowly along behind the dogs.

Luckily for me, all went smoothly, and we were soon approaching the cattle yards. Neville, the stock manager, and Nat, one of the shepherds, were busy weighing another mob of bulls. Neville glanced up at me as I approached. He stiffened, which caused Nat to glance up, too. The colour drained from their faces.

'Jeez, boss!' Neville said. 'What have you been up to?'

'Had a tangle with one of the bulls,' I said. 'Tossed me over a fence.'

Neville was a highly accomplished rodeo bull rider, and had seen his share of serious bull-inflicted injuries in his time.

'Shit! You'd better get it checked out at hospital,' he said firmly.

'Nah,' I replied, 'I'm right. You're over-reacting.'

He looked at me dubiously. Given his superior experience in being beaten up by bulls, this was probably a pretty ignorant call on my part.

I decided just to go home, wrap my leaky wounds as best I could in an old (but clean) sheet, and lie down for a while to let the effects of shock wear off. My only nod in the direction of being sensible was to ring a very good, caring friend and say that if I hadn't rung back in an hour, she should ring the ambulance and have it pick me up.

By sheer good fortune, it turned out I was right. My injuries were all superficial — cuts, grazes, deep gashes in my back and lots and lots of colourful bruising and gouges from the barbed wire up the backs of my legs, like stocking seams — and with a few days of rest and light jobs, I was on the mend. I later learned that it was one of our most excellent vets, Richard Hilson, who had driven past as I lay in the paddock. He told me he had been puzzled by the sight of my bike sitting on its stand with confused dogs and excited bulls wandering around the place. He had briefly wondered where I was, and now felt terribly guilty that he hadn't stopped to check the situation out. I would have been in good hands if he had. Vets are very skilled at practical first aid, and they're not bound by political correctness as proper human doctors are, which makes them a lot more efficient in times of major need. And lest there be any professional backlash or repercussions from this statement, all I should add is that their attention has the advantage of being tax-deductible!

It's not just large animals that pose a risk on the farm, and sometimes the dangers are difficult to foresee. Take the infamous possum in the blackberry incident, for example.

Blackberry can supply you with lots of juicy berries. But the birds love them as much as any connoisseur of apple and blackberry pies and, if left uncontrolled, the plant spreads very easily and can take over large tracts of land. The thorns catch woolly sheep and keep them captive until they are found or starve to death, so it can be a significant threat to production. The only effective means of controlling blackberry is to spray it, and the only way to remove it is to burn the dead remnants.

I had sprayed the blackberry bushes along the creek behind our woolshed, and by autumn they had died back to thorny loops like razor wire, perfectly dry and awaiting the right conditions to burn them off. One fine, cool day, a light southerly sprang up, which would carry the flames away from the buildings. I decided today was the day. I wound a piece of rubber inner-tube around a bit of pipe to make a blazing torch and I worked my way along the creek, setting fire to the dry, woody thickets of the pest plant.

Every so often, there would be a scrabbling noise and a shower of sparks as one of the many possums that lived in the blackberry took the fire as its cue to make a rapid exit. In some cases, when they weren't quite quick enough off the mark, they were well ablaze themselves by the time they burst into the open. I should have paid more attention to where they were seeking refuge, as when I got to the end of the creek and turned to look in satisfaction at the wall of blazing blackberry, I was horrified to see smoke coming out of the kitchen end of the old shearers' quarters.

By the time I got to the building, I could see that it was well alight up in the roof space. It was unlikely I could save it on my own. Luckily there was a cottage with a phone nearby, and the fire brigade were soon on their way. But I knew there was little chance they would get there in time to do much more than damp down the embers.

So it proved.

Despite the best efforts of everyone, the building was a write-off. To add insult to injury, we had to set it alight again at a later date before we could cart away the charcoal and twisted sheets of tin.

Bloody possums. The budget for blackberry control was never intended to stretch to a new shearer's quarters!

CHAPTER 9

HARD-HITTING LESSONS

It may be because I have been lucky with the stockmen I've employed, but I've had more dramas with machinery than with animals. Where stock is concerned, whenever I've had a tricky job on my hands I've been able to call upon someone who was already competent to help me out. This hasn't always been so with the machines. Due to the nature of our country, there has been no shortage of demanding mechanical work to be done, and when I've asked for volunteers to tackle these, the hands of the willing shooting into the air have been conspicuous by their absence. That has meant it's usually fallen to me.

Mostly, things have gone smoothly enough. But every now and again, we've had incidents, the majority of them situations in which the balance between gravity and traction has tilted onto the side of gravity.

Once, I was laying out the posts for an electric subdivision fence, a comparatively simple matter of driving the Land Cruiser along the line while someone on the back tossed a post off every 5 metres or whenever I called out. Well, it would have been simple, if it hadn't been for the fact that the fenceline traversed some steep country and it had been steadily raining for the past few days.

In those days we were letting one of our cottages to some beneficiaries who wanted a bit of a farm experience, and I had roped in the bloke who was in residence to ride on the deck of the Land Cruiser and do the actual throwing of the fenceposts. He wasn't a farmhand by instinct or calling, and seemed to struggle even with this most straightforward of tasks. I found I had to yell whenever I wanted a post deposited.

We reached a point where the line dropped down about 50 metres from a steepish ridge. It was nothing that would usually cause me any problems, especially since there was a good bit of easier ground at the bottom to serve as a run out if all went awry. So I confidently dropped over the lip, yelling out the window to my passenger to hang on tight.

You get a feel for these things. As soon as I was on the slope, I realised I had failed to allow for the combination of the extra weight of the fenceposts and the wetness of the grass, and that we were about to break traction sooner than I had expected. My focus shifted from feeding the posts out at convenient intervals to keeping the wheels turning at ground speed to maintain directional control of the vehicle.

As the Land Cruiser rapidly built up speed, I changed up gears to stop the wheels sliding and to keep it pointing at the run-out zone, turning my head slightly to yell to my passenger.

'Jump off!'

'Whaaaat?'

'Jump off!' I yelled again, injecting as much authority as I could into my voice.

I had done compulsory military training at school, and it was drummed into me there that when someone in command issues an order, you obey it at once and without delay. My passenger plainly had not had the benefit of any such conditioning. When I risked a glance backwards, I saw him frozen with fear, his knuckles white on the bars behind the cab.

I couldn't afford to worry about him any further. I changed up again, and might well have regained control had a very lethargic golden Labrador — Dylan — not been lolloping along dead ahead, totally oblivious to 3 tons of truck and payload rapidly closing in on him from uphill.

I tooted the horn, but Dylan flopped along untroubled.

I spied what I thought was just enough passing room before the hill got really steep. I aimed at it, only to find the hillside had slipped. The bit of hill that I was planning to drive on had vanished, and as we drove over the gap, gravity exceeded traction and we launched into mid-air.

Then the inevitable happened and the Land Cruiser started to roll.

I was no stranger to vehicular roll-overs: it's a common occurrence in 4WD rallying. I knew what to expect — although .

this time, I wasn't held securely in a four-point harness and cocooned in a full roll-cage, as I would have been in my rally vehicle. I wasn't even wearing my seatbelt. I braced myself as best I could against the back of the seat, gripping the steering wheel with one hand and the handbrake lever with the other. There was the distinctive booming, rending, crashing of bending metal and the crunching of glass, and all was confusion.

One roll isn't so bad. It's the subsequent rolls that are dangerous, as you inevitably lose your ability to hang on and begin to be bashed about the passenger compartment. And worse still, you can be flung through one of the apertures and crushed, which is how most fatalities occur in roll-overs of modern vehicles. Staying inside is the safest option, assuming it's within your power. Wearing a seatbelt helps.

I don't know how many times we rolled — it's hard to keep count of the number of times the darkness of earth and grass and the light of the sky replace each other in your vision. But after a few long seconds, the vehicle came to rest, thankfully back on its wheels. I did a swift check and found I was uninjured, apart from a few glass cuts and a lump on my head inflicted by a 10-inch crescent spanner that had been in the footwell on the passenger's side and had performed a high-speed tour of the cabin interior.

My thoughts were immediately for my passenger. The driver's door wouldn't open, so I crawled across to get out the passenger's side. I climbed shakily out. I heard moans. My passenger had completely ignored my instruction to jump off and had ridden the whole roll, clinging limpet-like to the deck. He was now lying over the side of the deck and was looking decidedly uncomfortable, probably because he had been momentarily pinned by the weight of the truck. He was spared the weight of the load of posts: they were scattered like matchsticks up the hillside from where the vehicle sat, crumpled, in the muddy fan at the foot of the slip.

The training in medical trauma from my old ski patrol days kicked in. I did the ABC test — Airways (he was making a fair bit of noise, so was plainly still breathing), Breathing (was happening, but looked painful) and Consciousness (he was fully with me) — and I decided that he was likely to have suffered internal injuries, possibly spinal damage. He needed expert help, and quickly.

Adrenaline kicked in, just as it had at the side of the doctor with the broken leg on the skifield eight years before, but I found I was able to summon the same calm, reassuring manner.

'I don't think you're seriously hurt,' I told him (trying to strike a balance between honesty and reassurance). 'We'll just get you comfortable and then I'll go and get a bit of help.'

I thought there was a danger that his breathing would be compromised if I left him dangling over the side of the deck as he was. But I knew that moving him would only complicate any spinal injury he might have. It was a hard decision, but strictly according to the hierarchy of human needs, I judged the need to breathe to be more important than the need to be able to walk or run or tap-dance. I lifted him as gently as possible onto the ground and laid him on his side.

The next step was to treat him for shock, which largely involves making your patient as warm as possible. This meant taking off most of my clothes and giving them to him.

Once I'd done what I could with the resources to hand, I had to decide what to do next. This was before the time of two-way radios. It was a Sunday, so there were no other staff about. I hadn't told anyone where I was going, so no one had the slightest idea where I was or what I was doing. No one would think to come looking for us for some time.

Forcing myself to breathe deeply and think clearly, I considered the available options. I was 8 kilometres from the homestead — too far on foot — but my cousin's house was only 4 kilometres away. That's where I would go to get help.

I explained to my patient what I was proposing to do, and as soon as he understood, I set off running (sort of) cross-country. It was a slippery scramble most of the way, with Dylan plunging through the quagmire ahead of me. Adrenaline kept me going, and I barely felt the cold. Every time fatigue began to set in, I thought about the man back there whose life might, even now, be ebbing away through internal bleeding. That would give me a new burst of speed. All the same, it was a long 4 kilometres, and I was relieved when I finally saw the house on the hill ahead of me.

My cousin had visitors, and they were visibly bemused as a part-naked man ran up their drive on a drizzly Sunday. But as I yelled for a paramedic and a helicopter, they soon got the picture.

I phoned the emergency operator and explained the situation. A frustrating conversation about the precise location of the accident ensued. I told them very clearly and precisely where it was, in the vicinity of Atua Hill, the second-highest hill in the district. They replied that they were looking at the map at the spot I was indicating but no such hill existed. After several precious minutes were wasted on this, they realised I meant Te Atua Hill, which was clearly marked, but then they proceeded to assure me that a conventional road ambulance would be dispatched. I didn't feel as though I ought to have to explain what it had taken to get a Land Cruiser to the site!

In those days, we didn't have a dedicated rescue helicopter service in the Hawke's Bay. A local farmer, Michael Groome, had a Jet Ranger fitted out for crop-spraying. I managed to get hold of him directly, and he set about stripping the spray gear out to make room for a patient. We arranged to meet at the Pukerangi airstrip, a hop and a skip from my cousin's house. A short time later, he reported that he had a paramedic aboard and was airborne.

For anyone who has been in an accident, the *wop-wop-wop* of an approaching helicopter is a very reassuring sound. After lighting a smoke signal with a bit of newspaper and wet hay, the pre-agreed indicator of where to pick me up, I waited as he set down. I was conscious of the time that had passed since I'd left my patient out there in the gully, and it was a profound relief to get the nod from Michael to approach the chopper, jump in and don a headset so I could direct him as the turbine whine scrolled up.

The ground dropped away below us, and in four minutes' flying we covered the route of my hour's adrenaline-fuelled cross-country scrambling.

The crushed-looking Cruiser was a sorry sight at the bottom of the hill, and so was the wretched form of the victim, lying there bundled up in my clothes. We set down as close as we could and the paramedic wasted no time in leaping out and starting his checks and the process of strapping the patient into a stretcher. A few neighbours, who had somehow heard what was unfolding, arrived by motorbike. Some of them had brought blankets, so I was able to recover most of my clothes. This was a good thing, as the adrenaline was wearing off and the shock and the chill drizzle meant I was experiencing violent fits of shivering.

Once the chopper had clattered off over the hill and I had warmed up a bit, it was time to work out what to do with the Land Cruiser. We certainly weren't going to get it towed from there, and there was no other easy option. It would have to be driven out.

We changed a tyre that had run off the rim, fitted chains to all four wheels, let some air out of the tyres (which has the effect of increasing traction), unbolted the collapsed driver's door (and in the process discovered that it had only been locked all along), and I wriggled inside, brushing piles of sparkling cubes of shattered glass off the driver's seat.

I tried the engine. After turning over for a few seconds, it fired and caught and the vehicle wreathed itself in reeking blue smoke as all the oil that had been flung into strange parts of the engine was cleared. I warmed it up, tightened my seatbelt as much as I could, surveyed the slip, picked a line back up and, applying every available scrap of horsepower, threw my heart to the top and launched into a seemingly impossible climb. Perhaps

155

my adrenaline had seeped into the engine, as the Land Cruiser seemed to perform like never before. It roared and bellowed and wriggled and slithered its way up the muddy channel, and cleared the climb with what the horsey set would call 'no time faults'.

As I limped into the turnaround area at home in the badly injured Land Cruiser, Chris came out. Her face was a picture!

'What the hell have you been up to?' she asked, aghast.

My hapless patient spent a few days in hospital. Thankfully his injuries were limited to bruising rather than the broken bones that had seemed more likely.

Lessons learned all round.

When I had built the deck for that vehicle, I had incorporated a well-braced roll-bar that I had fabricated from 50-millimetre hollow square-section steel, just like the one on the bulldozer. Although it had bent a bit, it had no doubt saved me from a far more serious outcome. Now I make sure seatbelts are worn tightly whenever I am driving in similar situations. I am able to tell classes in the 4WD training school that I run that I have rolled with seatbelts off, and rolled with seatbelts on, and much prefer rolling with seatbelts on.

The other sequel to the incident is that the paddock has worn the name 'Roller Coaster' ever since.

***

There was a time when tractor technology was very much less advanced than it is today, meaning that crawler tractors — which are, to all intents and purposes, bulldozers — were the only safe option for carting gear on steep, slippery hill country. Not that they were entirely safe. Once the tracks were clogged, as happens readily in our clay conditions, you might as well have been on skis. Trying to steer a crawler tractor skating backwards downhill with a trailer loaded high with posts is character-forming, to say the least. Your only option is to drop the blade with as much hydraulic down-force as possible to try to prevent a jack-knife.

The advent of 4WD tractors changed everything. Not only are they much faster, cheaper to run and more cost-efficient, but I find them safer, too, as when you're in a steep, slippery situation you get a bit of warning before you lose traction and start sliding backwards out of control. And even supposing you're being dragged by a trailer, as long as you're in 4WD and you've got tyres with the correct tread angle and at the correct pressure, you can control your descent — to a point.

I had been eyeing up an unproductive block, mostly comprising a steep, rough, broken slippery slope. I had decided that if I could disc it — turn over the soil with big cutting discs and prepare it for sowing with pasture grass — we could use it to break-feed the rising two-year, in-calf heifers. No one else was offering to do the job, and in fact one or two of the more proficient operators had pronounced one particular part of the slope as impossible to work with the discs. I didn't quite see it that way, and had worked out that if I were to turn the implement's cut off and drag it to the top of the slope by a different ridge, I could launch the tractor and discs off at the top and let gravity do a lot of the hard work.

So here I was, giving it a go. I sucked in my stomach and pulled the slack out of the seatbelt and then tipped the nose of the tractor over the brow of the hill, putting my feet up on the dashboard to brace. For a few minutes, all seemed to be going well — until there was an almighty crash behind me and a major thud on the roof of the cab. I brought the tractor to a stop, turned it off, and climbed out.

It was another of those 'if only I had a camera' moments! The discs were upended on the top of the tractor cab. The piece of number-8 wire through the bottom hole of the hitching pin had shore, and the pin had wriggled upwards and out. With the discs uncoupled, rather than simply staying put as the tractor went on without them, the drawbar had dug into the ground and the momentum of the implement had caused it to pole-vault onto the tractor. Luckily, I had a big tool bar on the three-point hydraulic linkage. That had taken most of the weight and possibly saved the cab from being flattened, with predictable consequences for the occupant (me). All the same, it was still a big mess.

I trudged home, planning the recovery as I went. I managed to manoeuvre my most capable 4WD vehicle — a Nissan Patrol in those days — close enough to the site to run the power-winch cable to the top of the discs. Applying the power, I toppled them back onto the ground, releasing the tractor.

After taking a crowbar to the mudguards to prise them off the tyres and using a sledgehammer to finesse out some of the dents, I re-hitched the discs, this time rigging up a double safety hitch and using extra safety chain. I finished the job with no further incident, and that paddock is now a picture, with high-quality green grass instead of rough old weeds and brown-top. It's a small block, but I would guess that it allows us to calve an extra five heifers per year.

Progress and profit — and never mind the cost!

I had another lucky escape around 2005, when I was using the tractor to carry a load of posts to fence a new block of land we'd recently bought from neighbours. Getting the posts onto the fenceline was not going to be an easy job, so I decided to do it myself. The 3-ton tip-trailer was loaded high with about 3 tons of 6-foot posts, and I had another half a ton or so of 9-foot strainers and coils of wire on the forks of the front-end loader.

I anticipated that the main difficulty would come at the point where I had to get the tractor and trailer across a tricky, steep-sided creek crossing. The plan was that the weight of the trailer coming down the hill would push the tractor through the boggy creek bed, and by the time the trailer hit the soft stuff, the front axle of the tractor would be on the hard ground on the far side. All that weight on the loader would serve to maximise the traction on the front wheels.

As I hit the creek, all six wheels (two on the front and dual wheels on each side at the back) clawing at the ground, I raised the loader so that the forks wouldn't dig into the far bank. I was concentrating fiercely on searching for traction, and failed to notice that as the front wheels climbed out of the bog, the loader was raised high in the air and the angle of the forks changed.

There was a rumble as half a ton of round strainer posts rolled over the back of the forks and down the frame to squash me. That got my attention.

There was nothing I could do, and before I could react the posts thudded into the front pair of the cab's four-post safety frame. The windscreen shattered, but the only injury I sustained was a few nicks from flying glass. My old John Deere tractor didn't have the substantial frame that the new machine did. If I'd performed the same stunt on the old tractor with its two-post safety frame, you wouldn't be reading this.

After I had loaded the posts back on by hand, the job was completed and man and machine were safely home again, I reviewed the situation, and added a pair of safety bars to the loader forks that would intercept a load rolling backwards, as this one had. I've never seen this addition on another loader. Perhaps I should patent it!

As I've said, blackberry has a lot to answer for. Some farmers hire a helicopter to spray it from the air, but we always did it on the cheap, mostly using a spray rig on the back of the tractor.

One day, I was out spraying blackberry on our John Deere 1040 with two 44-gallon drums of pre-mixed spray strapped to the transport tray on the three-point linkage behind the machine. The aim was to get to within 50 metres of each blackberry bush, so that the spray-gun could be dragged to the end of its 50-metre hose and the bush spot-sprayed.

I was operating the tractor in reasonably steep country on the edge of a paddock called Pond Hill. When climbing off, I was always very careful to pull the handbrake on as hard as possible and to turn the front wheels into the hill to prevent a premature, unmanned departure down the hill. Climbing back on board, I was just as careful to pull my seatbelt on as tight as possible before I set off back down the hill.

Good thing I was meticulous about these precautions, because just after I had set off this time, the tractor suddenly shot forwards down the hill, crashing and banging over the rutted ground. My teeth rattled in my head as it bucked and bounced. I briefly considered jumping off, but with the extra width of the dual wheels (fitted to make the tractor more stable and to get better traction on steep hills), I doubted I would avoid being run over. With the situation deteriorating very rapidly, I noticed that the front wheels weren't sliding, which meant they were disengaged. The 4WD system on this tractor relied upon a complicated actuation process that involved an electrical switch operating a hydraulic mechanism that manoeuvred the mechanical linkages into position, and it had already proven prone to malfunction. I quickly flicked the switch off and back on, and almost immediately the front wheels got some grip. By manipulating the speed of the wheels, I regained control. The other outcome would have been very messy. The bucking and bouncing had left a trail of tools from the mudguard-mounted toolbox scattered down the hillside, and it was only a matter of time before that bucking and bouncing threw the machine off-line and it rolled.

Another memorable blackberry spraying incident happened when my son Henry and I were spraying blackberry in the Kopere Block. I had driven the new Deutz tractor as far as it would go up a steep gully and Henry was lugging the spray-hose up the rest of the rocky cliff to douse a bush that we could see perched up there. He was only 12, and the hose weighs a bit. He was struggling to climb and pull it after him, so I thought I would climb behind him and take most of the weight of the hose. I hopped off the tractor and grabbed the hose halfway to help.

I was concentrating on clambering up the rocky face when I heard a blood-curdling scream from Henry.

Fearing the worst, I glanced up at him, just as I felt an insistent tug on the hose from behind me. Henry was throwing the gun as far as possible out from the cliff face to stop the gun being damaged as it dragged through the rocks.

'Dad!' he yelled. 'The tractor! Going down!'

The hose was snatched from my hand. I turned to see our shiny new tractor gathering speed back down the gully, the spray-hose slithering behind it. The tractor rolled backwards down the slope but, instead of running all the way down to almost certain destruction, it veered across the valley and ran up the other side, where it bellied out on a rock.

'Are you OK?' I asked Henry, who was as white as a sheet.

'Yes,' he said. 'But what about the tractor?'

It turned out that the main damage was to the bottom of our new 600-litre spray tank. The chemical solution was trickling out of the ruptured tank. We had about 400 litres aboard, about the value of a good cattle beast, and I was buggered if I was going to stand idly by and watch it go to waste. We went into problem-solving mode. I swiftly calculated the cubic capacity of the bucket of the front-end loader, and decided that if we disconnected it from the front of the tractor and propped it up with rocks, it would serve as an emergency holding tank.

We quickly unhitched the bucket, and, while Henry wedged stones at strategic points beneath it, I managed to manoeuvre the tank over the top of it. We collected just about all of the spray, and then it became a simple matter of driving the tractor the 8 kilometres home, repairing the tank and returning the next day to refill it with a hand bucket. We got the job done, this time using a backpack sprayer to reach the trickier bushes.

It turns out that bulldozers, too, have a stability limit. I found this out the hard way when I was replacing one of our conventional fences, which required me to somehow wrangle a load of about 60 posts and coils of wire over one of the roughest blocks on the station. The only feasible way to do it, other than laboriously by packhorse, was to haul it up a heavily scoured river bed using our BT6 bulldozer and dual-wheel trailer. The theory was that I could use the blade as I went to make a bit of a track.

With the 2-ton weight of a post-encrusted trailer on behind, the bulldozer developed an encouragingly increased pushing ability downhill, handy when cutting a track down a 10-foot bank into the stream bed below. But by about the third crossing, I might have been getting a bit overly optimistic about its new-found bank-displacing ability. I struck a particularly hard bit of dirt, and decided I needed to back up for another chew at the bank. But the weight of the trailer, already over the brink of the hill behind me, refused to allow any degree of pushback. The tracks clawed unsuccessfully at the mud as I tried to reverse, and this had the effect of clogging them and drastically reducing traction. Gravity took over. The dozer slid sideways, jack-knifing the trailer, and when the tracks encountered a hard, flat rock, the whole ensemble slid sideways down the bank into the creek. It was all over in a blur, and I found myself still sitting in the driver's seat, but at a 60-degree angle to the horizontal and with about 2 tons of posts perched above me, the strops holding them onto the trailer creaking and pinging.

Recognising that if the bottom track dug in any further it would cause the machine to go completely over, I backed off the power. I jumped off into the creek — contrary to your instincts, it's easier and actually safer to jump off on the downhill side of a rolling vehicle — and struggled off to a vantage from which to survey the situation from a safe distance. Looking at the predicament, it was easy to accept that getting the machine out of the boggy mess without help would be all but impossible. There is a certain point at which it pays to admit defeat. Climbing gingerly back aboard, I pulled the stop control and jumped with a squelch back into the mud to begin the frustrating, 4-kilometre walk home, planning the complicated recovery as I went.

Luckily, Clareinch Station next door also had a bulldozer, albeit one a bit smaller than ours. Stewart Kingston, the manager, came to the rescue. I mounted our 4WD tractor and we drove in convoy to the scene. Stewart nearly managed to keep a straight face when he saw the mess, and we put my plan into action. We secured a chain to the uphill track frame of my dozer, and Stewart's machine took up the slack to stop it tipping over completely. I attached another chain to the rear of the trailer and gently took up the strain with the tractor. I used an oxyacetylene torch to cut the drawbar pin securing the trailer to the dozer, and then used the tractor to drag the trailer clear.

After overfilling the dozer engine sump with oil (to ensure the oil pump wouldn't be sucking air, given the radical angle the engine found itself on), we started her up. Stewart and I both agreed it was too dangerous to operate the dozer from the driver's seat, so I attached a few strands of baling twine to the clutch lever and, using the wheel tractor to pull the blade forward and the Clareinch dozer as an anchor, I stood a few feet clear up the hillside and operated the clutch with the twine. In this way, we were able to coax it from its precarious position.

I then did what I probably should have done in the first place. With the dozer unencumbered by the trailer, I went ahead and built a track up the river before returning and reconnecting the trailer with a new drawbar pin. Carting out the posts was then accomplished relatively simply on the new track.

Once it was all over, the realisation of what might have happened hit home. The incident preceded my marriage, so when I got home at the end of the long, scary day, there was no one to talk it through with. Rum helped, but not for long. Reflecting on what might have been, I put the dozer in the workshop and spent the weekend building a roll-bar that I welded up out of

100 x 100-millimetre hollow steel. I also fitted a seatbelt. I had got away with a severe lesson, but at least I had learned from it.

***

It's amazing how the mismanagement of machinery can complicate simple jobs. The Waipari bulldozer was pretty elderly when I took over the farm, but as I didn't have a lot of spare cash to throw at maintenance, I worked around its foibles as best I could. The batteries were well past their prime, for example, but I couldn't afford new ones. They had enough power to heat the glow plugs — necessary to ignite the first charge of diesel in a simple compression engine — but not enough to crank the engine over. The solution was to park the dozer at the top of a hill with a short post under the edge of the blade (the seals in the hydraulic rams were perished, and the blade would settle overnight if you didn't prop it up), so that when you climbed aboard, you could heat the glow plugs, select third gear, disengage the brakes and let the machine gather momentum downhill. Once you were going fast enough, you let the clutch out and, all going well, the old motor would fire up with a series of reports and the production of two or three impressive white smoke rings from the stack.

One morning, I arrived to do a quick job with the dozer. We were doing a big clean-up of some old macrocarpa trees, and had salvaged the best logs to mill and were burning the rest in a great heap. This required using the dozer to push the burning heap back together each morning to ensure everything was consumed.

I had plenty of other work to be getting on with, but this would take half an hour, max.

I kicked the post out from under the blade, climbed aboard, twisted the key until the red indicator showed the glow plugs were to temperature, shoved it into gear and stabbed the brake pedals to disengage the brakes.

Nothing happened.

I swore. It was clear the hill wasn't steep enough.

I jumped off and found an old post, which I wedged in at the heel of one of the tracks in an attempt to lever the whole thing into motion, the plan being that as soon as it began to roll, I would scramble aboard before it gained too much momentum. I heaved and strained and swore in vain. It wouldn't budge.

I went and got the wheel tractor, with a view to backing up against the dozer and giving it a nudge to start it off downhill. Realising that I wouldn't have time to stop the tractor, apply the brakes and jump off before the dozer had started careering downhill, I tied a stout, 5-metre rope between the two. This, I judged, would hold the dozer for long enough to allow me to get aboard and apply the brakes, whereupon I could untie the rope, climb back aboard, crash-start as usual and go on my merry way.

It didn't quite go according to plan.

The dozer got going more quickly than I had calculated, and as I hurried to take up the slack in the rope, I accidentally selected too high a gear and stalled the tractor. Suddenly, I was back in sixth form physics, with Hoss Harwood leaning over me.

'It's quite simple, boy,' he was growling. 'Mass times distance times speed. Think about it!'

The answer came to me far faster than it ever had back in the day. Trying to stop a 5-ton bulldozer from careering driverless down a 50-metre hill with a 3-ton stationary tractor wasn't going to fly.

Jumping off a tractor is never a satisfactory choice for me. In this case, it was unlikely I could safely jump clear. And it would leave two expensive machines piled up at the bottom of the hill instead of one. So that was out. But backing quickly downhill behind the runaway dozer in an effort to slow it gradually seemed a risky proposition. Beyond a certain point, jumping off really would be out of the question, and the alternative was ending up amongst the wreckage of the two expensive machines.

My choice was clear. Hanging on tight, I dropped the blade on the rear of the tractor and jammed on the brakes as hard as possible, hoping that the dead weight would slow the dozer down enough at the bottom of the hill to make for a slightly gentler crash.

The dozer gathered speed, the rope came up taut and there was a loud bang and a jolt as, perhaps luckily for me, it snapped. The tractor stayed put as the bulldozer found its natural way into a hole in the creek at the foot of the hill, where it sat smugly until I could get the tractor alongside, start it with jumper leads and drive it out. I could have started it with jumper leads in the first place. Instead, I had wasted half a morning.

More hard lessons!

The closest I have ever come to death was when I was bulldozing a new track down our Kopere Block, a 67-hectare patch of land right out the back of Waipari Station that we used for calving. Like much of Waipari — and in fact, like much of coastal Hawke's Bay — Kopere is pretty steep. When I took over the farm, it could only be worked on foot or by horse. But I had a dream of making it so that the whole station, this block included, could be mustered by motorbike, perhaps even driven through by 4WD. So one spring I embarked on an ambitious plan to push a 3-kilometre track down the steep face and through the valley below.

As usual, I was doing it myself in order to cut costs, using the station's BTD6 bulldozer.

Some of the pushing was reasonably easy, being silty clay. When you're pushing this stuff down a steep downhill, it's fairly easy to make good headway. But the problems began when I made to carve a hairpin bend and struck solid limestone rock. This was much harder, and to make any impression on it you had to give it the charge with the blade, chipping it away bit by bit, with sparks and smoke coming off the blade at times. It was bloody slow work. I wished I had a few sticks of gelignite handy, but I didn't, and finding some would have wasted more time. Besides, I wasn't sure I could remember the right combination of fuse, cordite, detonator and gelly from my ski patrol avalanche-blasting days. So I was stuck with the bulldozer. One 5-metre section took me all of a 10-hour day to cut through. Plus it was very hard on gear. At one point, I broke off the 30-millimetre high-tensile bolts that secured the blade frame to the dozer itself, which meant I had to cart a gas cutting-and-welding rig and a generator up there so that I could drill out the old bits of bolt, re-tap the hole in a larger-size thread, and reattach the sub-frame. None of the tools was perfect for the job, so it was an exercise in frustration. It wasn't just the machine that was feeling the strain.

The work was repetitive. I was perched on the 3-metre-wide track I'd carved out of the hillside. The target area was on the outside of the track, where the level area met the original hillside. I would select forward gear, pull on the clutch control and concentrate on keeping the blade digging into the rock at just the right depth — not so shallow that the blade deflects off without moving any material, and not so deep that it digs in and the tracks start scrabbling uselessly. With each gouge out of the rock, I'd have to ensure the blade was set at such an angle that the rubble was shunted across the front of the blade to where it would rattle its way down the 20-metre drop to a creek below. The dozer was 5 metres long. To perform the operation, I had to work at a 45-degree angle to the track, and ensure I didn't hang the tracks too far over the edge on each forward pass, or hit the inside bank as I reversed up for another bite. At the end of each nibble, I'd have to select reverse gear, engage the hand clutch lever, lift the blade with the hydraulic control, and pull one or both of the hand-steering levers to line up the target.

It was late afternoon, and I'd had hours of this. So perhaps it was fatigue or an inability to concentrate on the repetitive task at hand (my mind was often miles away planning another way to tackle a difficult task). Or perhaps it was the way my brain works (or doesn't) sometimes — I'm dyslexic, and every so often I'll find myself doing the exact opposite of what I intended to do.

Whatever it was, I missed an essential part of the sequence and, with the motor roaring at full power, looking behind me to pick a reversing line, I pulled the hand clutch firmly back. Instead of moving backwards toward the inside bank, the dozer shot forwards, towards the edge. I reacted at once, and in a split-second I had dropped the blade, hit the clutch and jammed on the foot brakes.

It's at moments like these — and I've experienced a few — that you become acutely conscious of gravity. The dozer teetered, 5 tons of steel rocking with almost gymnastic poise and grace on the edge of the track, with the engine bellowing and the 20-metre drop beckoning to me, and my life flashing before my eyes.

As the dozer tipped back and forth gently as though trying to decide between life and death, I throttled the engine back to an idle. I sat for a minute or so as I calmed down, then pulled the fuel cut-out control, locked the brakes, turned off the key, climbed off and walked away.

It was a few days before I got my nerve back, and even then, the sight of the dozer poised there gave me pause for thought. But I executed the recovery plan I'd formulated late at night, in between bouts of fitful sleep and dreams of falling. I positioned some old posts under the blade to give me a solid platform against which to lever it, which had the effect of pushing most of the dozer's weight onto the rear part of its tracks. Once this was done, I was able to gently ease the machine back onto hard, safe ground. I was lucky I had been working in rock rather than in the more usual clay — it gave a fairly hard footing and didn't give way too easily.

Whenever I pass that bit of track I still shiver at the thought of what may have happened, although the track itself has, as intended, transformed the way we work that bit of country.

I realised, with the benefit of hindsight, that I ought to have got a bigger machine equipped with rippers on the job on that part of the track. When later on I put in another similar track, I employed an experienced contractor, Ken Scheele, who used his Caterpillar D5 equipped with rippers. Ken was a legend in the district, and was still driving his bulldozer at 85. But he didn't come cheap: I'm not sure which gave me the bigger fright — my near-death experience with the bulldozer or Ken's bill! With Ken you sometimes also had to factor in the costs of new gateposts, as he seemed determined to get an 11-foot blade through a 10-foot gateway.

CHAPTER 10

THE SECOND MOUSE GETS THE CHEESE

I first went to a Hawke's Bay Farmer of the Year field day in 1985. Each year, around 10 top-performing farmers from the district are 'encouraged' to enter the competition, which is organised by the Agricultural and Pastoral Association and seeks, as the 1986 winner, Sam Robinson put it, to identify a farmer and a piece of land who together showcase farming excellence. It's one of the most interesting and often exciting days in the Hawke's Bay farming calendar.

The 1985 field day was on Bruce Worsnop's property at Tikokino and, Bruce being a bit of a showman, it was always going to be an entertaining day out. But I realised at once that the FOY is also an excellent opportunity to hear about and observe the practices of others. Various financial figures are analysed, with the overall indicator being the farm's return on capital, a number that provides a realistic comparison of performance across different types and sizes of farm. Over the next few years, the FOY field day was an unmissable event for me, as I absorbed the wisdom and tricks of the very best farmers and attempted to implement aspects of their wisdom at Waipari. Not for me the gossiping that goes on at the back of the crowd; I was always front and centre, hanging on every wise word, asking copious questions and using my trusty, solar-powered pocket calculator to crunch the numbers to get a feel for how Waipari was doing when compared to the best performers.

It wasn't all work, though. A centrepiece of the day was the farm tour, when about 300 people would be conducted around the property in the interiors and on the decks of about 50 4WD farm vehicles. In fact, this part of the day became a sort of unofficial competition in itself, to determine who had the best and best-equipped 4WD vehicle. In 1985, fittingly, Bruce was the clear winner, with his brand-new V8 Land Rover, kitted out with the biggest and best mud tyres money could buy. Bruce still hedged his bets, retaining his oversized Chevy Silverado pick-up as the showpiece, with Kate, his wife, driving the Land Rover with their young twins, Patrick and James, inside.

Inevitably, someone was bound to get stuck, and that would be the cue for a good deal of good-natured ribbing from the rest.

By about 1990, Waipari seemed to have well and truly turned the corner, and when I sat down with Roy Fraser, a well-respected farm adviser, to look at our numbers compared with those of the Farmer of the Year, I realised we had caught up a bit. So with Roy's encouragement, I put in an entry in 1991.

To my surprise and delight, Waipari made it through to the final few. I had presented the judges, Tom Atchison and Jim Scotland, with a comprehensive computer printout of comparative gross margins on the various livestock classes, and this demonstrated that some of our unorthodox livestock policies were generating up to three times the Hawke's Bay average economic farm surplus (EFS) per head (or per stock unit, as we had become accustomed to saying). Even more impressively, the figures also indicated that the following year's performance would be better still. I thought we might be in with a chance.

Tom and Jim came out to the farm to do an inspection, and I loaded them into the Land Cruiser to do a bit of a tour. It had been raining a fair bit, so I had taken the precaution of putting chains on all four wheels to help with traction on the slippery bits. From my perspective, it was all going smoothly, and I was happily showing off all my favourite achievements when Tom asked me how I thought we could safely run a FOY field day tour on Waipari.

'What do you mean?' I asked, surprised at the question.

'Well,' he said, 'don't take this as a criticism of your driving or anything — I feel perfectly safe driving with you. But we did just slide that last 200 metres down the laneway even though we've got chains on all four wheels.'

'Oh, it wasn't that bad, was it?' I scoffed. 'It was all controlled.'

The silence from the esteemed judges was deafening.

The trip home was uneventful, and we exchanged lots of small talk and pleasantries. But I had a hollow feeling when they left. Chris tried to brighten my mood by telling me that, if nothing else, being in the running for FOY had got me to tidy all the boots at the back door for the first time she could remember, so all had not been in vain.

A few weeks later, I went along to the field day on the farm belonging to that year's winner, Chris King of Onga Onga. It was a cropping farm, and couldn't have been more different to Waipari — flat, tame and dry. I wasn't letting on to anyone that I had been judged but beaten. We were standing listening to Tom Atchison's speech, in which he was singing the praises of the cropping farm as a very efficient and profitable enterprise.

Tom paused.

'Still,' he said, 'no one should be too quick to write off the traditional hill-country sheep and beef operation. My fellow judge and I recently judged a large, rough coastal station that was a top performer, and right up there, in our opinion.'

Because the Onga Onga farm had little in common with Waipari and not so much to teach me, I was at the back of the crowd for once. I thought I would burst with pride!

Later on, at the after-match function, I came across a mate, Steve, whom I knew had also entered. He was having a sad, quiet moment to himself in the bushes. Despite the affirmation I'd just received, I knew how he felt. So much time, effort, heart and soul goes into farming that when you fail to receive the recognition you feel you're due, it can come as a bit of a kick in the guts. I did my best to console him. In the end, I don't think it was my lame, mumbled words but rather Steve's own sheer determination that saw him win the award a few years later, and several other farming awards besides.

Naturally enough, given the encouragement I'd received in 1990, I re-entered the following year, this time putting a fair bit of effort into making sure the judges didn't get scared off. Tom had retired by rotation, and Gerry Sainsbury, the inaugural winner of the award in 1972, had taken over from him. The forecasts we'd prepared in 1990 had been fully vindicated, and we'd had a good year. It helped, too, that when Gerry and Jim came out to do a site visit, it was a much drier day and we were able to get around the farm without anything approaching the excitement of the year before.

One evening a few days later, the phone rang.

'Mark Warren? Bill Crawford here.'

I went cold. Bill was Chairman of the Hawke's Bay A&P Association Farmer of the Year Committee.

'Congratulations, mate,' he said. 'You're Hawke's Bay Farmer of the Year.'

You don't forget moments like that. Receiving that call is right up there with some of the best moments in my life. I put the phone down, and the yahoo was loud and long. The phone rang again, with the first of many people who were in the know calling to offer their congratulations. The feeling was the opposite of what poor old Steve had experienced the year before; all the hard work, the long hours, the anxiety, the risks and the heartbreaks seemed to have paid off.

At the earliest opportunity, I rang my parents to tell them, in as modest a way as possible, that there was going to be a field day at Waipari for the Farmer of the Year.

'Oh, that's interesting,' Dad said. He had never been particularly ambitious, or that interested in material progress, so focused was he upon his life and work as a vicar. He didn't really get what the farm involved.

It had always bugged me that whenever I had reported some little triumph or another from Waipari to Mum and Dad, they had taken the opportunity to remind me that, without help from them, I couldn't have gone farming in the first place. But this award was judged on facts, without any input from emotional opinion. It was great to be able to produce objective evidence, from those whose knowledge counted, that I had been doing something right, without having to rely on my family background. I had proof that all my hard work and, I hoped, my abilities, were making a difference.

Others were more generous. Pita Alexander, our very good specialist farm accountant, wrote me a letter filled with wise words about how to make the big day as good as possible. He had watched and guided as I turned Waipari around, and much of what he wrote showed that he probably knew me better than my own father did.

'By nature, you are an impatient animal,' he wrote. 'You don't have much time for duffers, and don't bear fools lightly.'

Sam and I spent weeks planning stock rotation so that the best of the stock would just happen to be beside the laneway on the day of the big event, and all the tracks were made as safe as possible, which required installing safety ropes at one potentially dodgy place coming down through the bush. But strictly according to Murphy's Law, at two in the morning of the field day, the heavens opened.

Chris and I lay there listening to the rain hammering on the roof until four, when we got up and set about implementing Plan B. We set to and built huge photo-boards of the history of Waipari's development.

I still held out hope, and set off at first light on the 23-kilometre round trip to check the track along the tour route. I managed everything easily with just one set of chains on the Nissan Patrol. I arrived back at the shed just before 8:30am, convinced that it was still safe for the 300-odd guest farmers to travel around the station. But several of my mates had arrived to help set everything up. I could see them standing in the door to the shed looking through the rain. They were looking at the mud-spattered Patrol and its filthy windscreen, and shaking their heads. Peter Butler and Bruce Worsnop took me aside for a wee chat and explained that it would be foolhardy to expect the multitude of farmers to drive over the ground I had just covered without there being a serious accident. With a heavy heart, I accepted the advice of my elders and betters. We agreed we had to cancel the best part of the day.

Once the crowd had assembled, however, and after the preliminary greetings and briefings, we set off in convoy for the short drive up onto the coast block, which allowed a brief glimpse through the misty rain of what they might have travelled over. The sight of over 60 quad bikes snaking their way up the muddy hill had to be seen to be believed. Besides the bikes, there were a couple of 4WD tractors with trailers provided by neighbours, serving as hard-seat, open-air buses, while the VIPs were chauffeured in proper 4WDs with chains on all four wheels. I had roped in my good friend Hugh Williams to look after Mum, driving her on the tour in his chained-up Hilux. Dad couldn't be there on the day.

The highest point on the coast block gave a view of much of the rest of the station. It gave people some sense of the challenges we faced working Waipari. But we couldn't venture further afield in those conditions, so we all returned to the woolshed, where the organising committee decided we would spend the rest of the day. Lucky it's a big shed! As the rain drummed incessantly on the roofs, I was duly elevated to the podium alongside Bill, and spent the rest of the afternoon telling everyone what we had done in order to achieve a 15.9 per cent return on capital. There was much lively debate and even some who openly doubted our figures, but the day seemed to deliver on the objective of sharing quality farm-management information. Meanwhile, Sam ventured out on a quad in his wet-weather gear to open the gates to the various holding paddocks where our best-looking stock had been waiting patiently in the rain for an appreciative audience that never came.

I was worn out but on a high after most of the guests had left, but it wasn't over yet. The title of Hawke's Bay Farmer of the Year came with a cheque for $5000, and I had decided I would spend it on a big party to thank all those who had helped us turn a loss-making business into a very profitable one — and now pick up FOY along the way. The party was a ripper that went long into the night, eventually retiring to the homestead. And from memory, Sam, who was usually a very shy and retiring type, did a legendary performance of his party song, 'I wish I was an apple / a-hangin' in a tree!'

Not that being Farmer of the Year was all beer and skittles. A few weeks later, when I was in town trying to buy some farm equipment, I asked as usual for a sharp price if I paid cash up front.

'You don't need a discount from me,' the dealer snorted. 'The papers reckon you're making plenty already.'

Back to reality!

<hr>

Part of picking up the title of Farmer of the Year was the outside interest that was generated in the way we did things on Waipari. It was both flattering and encouraging to have other, well-established farmers ring me to ask questions about how I did some of the things that I did differently, especially when I looked over their fences some months later and saw, in some cases, evidence that they had changed their farming policies to align with mine. The media were interested, too, with the phone ringing constantly with requests for interviews on radio and in the papers. Not all the commentary was favourable. Some of my unorthodox thinking seemed too left-field for some people — proof, perhaps, that dyslexia can have its advantages.

One day, I received a phone call from Julian O'Brien, who was director of the long-running TV show *Country Calendar*. He wanted to come and shoot a show on profitable hill-country farming.

I was chuffed, of course, and agreed.

Julian came up about a week later for the exploratory interview, which consisted of an hour or so of questions as to how they might make a programme. I took him on a quick trip up to the coast block in the Land Cruiser so that he could get a bit of an overview of the station. As it was a wet day, the track was a little skatey, and a bit of 'spirited driving' was needed to get up the hill. Julian swore a bit on the way up as the Cruiser slithered and clawed its way noisily up the track, and on the way down his curses seemed to be interspersed with snatches of prayer, but I thought that was just the film industry way. It wasn't until a year after our show was done and dusted that I learned differently.

Julian was filming a programme about the legendary helicopter pilot Dave Saxton when the chopper engine cut, forcing Dave to auto-rotate it into an emergency landing. They all walked out OK, but on hearing about the incident on the TV news, I rang Julian to offer my sympathies. 'Must have been a bit scary?' I said.

'Fuck me days,' he replied, 'it's the only time I've been more scared than I was out on the coast in your bloody Land Cruiser!'

I thought that was a bit harsh on my driving!

But anyway, the trip up the slippery track and back notwithstanding, Julian must have seen potential for a show, because the *Country Calendar* film crew — director, cameraman and sound man — duly turned up to shoot on Waipari. Doubtless inspired by Julian's recent experience, the first sequence was to show the Land Cruiser with chains on all four wheels carting fencing gear out to a new development block. I was instructed to try to outdo 'Crumpy' — Barry Crump, as featured driving recklessly in a legendary TV advertisement for the Toyota Hilux — and I was confident I could do a bit better than that. For some reason I can't recall, the exhaust system had been torn off the Cruiser and it was running a straight free-flow exhaust. And it wasn't all sound and fury, either: the motor had been considerably hotted up with tricks I had learned from my days building my rallying Land Cruiser, and it was decidedly more sprightly than your normal 155-hp farm truck. The resulting film footage — in which I wallowed across deep, muddy bogs with a load of posts and assorted paraphernalia bouncing around on the tray as I kept the four chained-up wheels chewing wildly, all to the accompaniment of the unmuffled roar of a full-revving engine — made for a great opening shot.

A week's hilarity and high entertainment followed, with the crew attempting to film some of the more exciting parts of everyday, hard, hill-county farming. One shot called for me to run the giant discs over a rough hill development block. The conditions were far from ideal — it was wet, and we deliberately chose an unpromisingly steep tract of land — and I may have pushed my luck a bit far. Despite dual wheels on the 4WD tractor, I couldn't recover its position. I had to go and get the bulldozer to pull it out of the boggy hole, and because there was only one experienced dozer driver on set (me), I had to give the director a few quick lessons on how to handle a BTD6.

The film crew thought this was all gold, but as we were possibly breaking just about every OSH rule in the book, I regretfully asked them not to use the shot. They decided they would have to settle for using it as classified footage to be screened at the TV company's Christmas party.

Being mid-winter, rain and mud were constant companions. With one shoot to go before it was a wrap, the rain seemed to get into the $80,000 German camera and it jammed. By now, after a full week spent filming, I was getting a bit behind in my real farm work, and didn't have time to wait for a new camera to be flown up. So with the crew hovering nervously, I did what I'd do with any other bit of machinery and took to it with a teaspoon and pocket knife on the kitchen table. After a few minutes tinkering, while the cameraman stood looking helplessly on like the parent of an infant undergoing open-heart surgery, I was proud to announce I had got the thing going again. That afternoon, between showers, they filmed me mustering in calf, rising two-year-old heifers with my four-year-old daughter, Emma, seated in front of me on the tank of the farm bike, wearing her pony helmet (suggested at the last minute as a nod in the direction of safety by Chris, her mum). Emma held on stoically as we bounced down the muddy, rutted hill in the pouring rain. Once at the bottom, I lifted her off the bike to finish putting the heifers through the gate. The footage captured Emma reaching up to slip her hand in mine, as she always did when we got off the bike, a gesture that indicated that all this scary-looking stuff was old and familiar territory for her. It was a brilliant cameo performance!

Soon after the show went to air, the comments started rolling in. Many — perhaps even most — were highly congratulatory, but some were not so enthusiastic.

'What did you think you were doing,' someone thundered, probably from their couch in a warm city apartment, 'dragging a child that age out on a farm in those conditions?'

People in the city just don't appreciate what difficult, uncomfortable and occasionally dangerous lives farmers and their families live. But it wasn't just city-slickers. I also got a call from the father of a farming mate of mine in North Canterbury.

'You're nothing but a bloody hoon,' he said, 'the way you were carrying on in that vehicle.'

I thought about my mate, his son, misjudging a river crossing in their station ex-Army Unimog and making it across with muddy river water washing halfway up the windscreen, then drily remarking that it had all got a bit untidy. I thought about him running a tyre off the rim when he was doing circle work in a rental car on the beach, and the mad scramble that followed to change the wheel before the incoming tide submerged the vehicle. And there were plenty more stories where those came from.

I chose to say nothing.

Letters rolled in. There was one from the *Country Calendar* team, thanking me profusely for making the show happen (although they couldn't refrain from bemoaning the $8000 worth of damage I seemed to have done getting the soggy camera to go again). But the one that meant the most to me was the congratulatory letter that I received from my old headmaster at Waihi, Peter Prosser. Peter recalled that my focus seemed to have been more firmly fixed on the Mount Peel range out the window than on my schoolwork, and that he had known from the outset that I was unlikely to follow my father and grandfather down the academic route into theology and the clergy. But he told me he had watched the *Country Calendar* programme on Waipari with a mixture of surprise and admiration. He hinted that in light of the farming success I had achieved, my previous scholastic failures and inadequacies were forgiven. Unlikely as it had seemed back at Waihi, he said, it looked as though I had actually managed to make something useful of myself.

As my mate Steve (the unsuccessful nominee for FOY in 1990 who had subsequently made good) would no doubt agree, recognition may be slow to come your way, but once you receive it, it all seems to come at once. In the wake of all the media attention I got, I was delighted to be nominated for the AC Cameron Award for Excellence in Farming in New Zealand, and while I didn't win it I did pick up a silver medal as winner of the eastern districts, a very pleasant surprise. And in the following summer (1993), I was inducted into the Kellogg Rural Leadership Programme at Lincoln University. My Grandfather Alwyn had always insisted I do my bit for Federated Farmers, and, although I had already been nominated as chairman of the local district branch (Omakere) of Federated Farmers, I felt that if I were to take on further responsibilities in farmer politics I needed to be both a well-proven, successful farmer and to have been trained in such un-agricultural skillsets as how to deal with the media. There used to be a perception that Federated Farmers was just a retirement option for former farmers: I certainly didn't want to add to it. The inspirational six-month Kellogg course served this purpose very well, and I made many good friends and contacts besides.

But it turned out that the downside of receiving a bit of training in public affairs is that you then come under constant pressure to take on community service jobs. You learn quickly that you have to be firm and say 'no', otherwise you'll soon find yourself spread pretty thin.

Even so, at the start of 1994 I seemed to have risen to the position of Chairman of the Meat and Wool Section and Vice-President of Hawke's Bay Federated Farmers, more because I was the only willing candidate than because I prevailed in tight-fought contests for the jobs. I soon discovered why.

One Friday afternoon, after a fairly humdrum stint in the Federated Farmers office, I wandered over the road to the Hastings Club for an evening refreshment. No sooner had I walked into the bar than I got talking to one of the club stalwarts, Fenton Kelly.

'How's it going, being our Meat and Wool chairman?' he asked.

'Pretty cruisy so far,' I shrugged. 'No major dramas.'

Fenton took a considered sip of his gin and tonic.

'That might be about to change,' he said. 'Have you seen the news?'

When I told him I hadn't, he broke it to me. Weddel Crown had just gone into receivership. They owned the big Hawke's Bay meat works at Tomoana, which, besides being a major employer, was also a significant outlet for North Island livestock.

No sooner had I learned of this development than my new-fangled Motorola cellphone, a brick of a thing by today's standards, and worth 20 good lambs, began ringing. It didn't stop for the next few days. Farmers were anxious to know how they could retrieve from the corporate maw the stock they had sent for processing but had not yet been killed, and at what point the stock became the property of the receiver rather than the farmer. Many also wanted to know how such a large and apparently profitable operation could fall over without warning.

These were all good questions, and over the next few weeks we did our best to find answers and also to salvage what we could for farmers. I was one of a small investigative team that was allowed into the Weddel Crown office to inspect their records in order to work out who owned what. While I was there, I used my flash new mobile phone to communicate with colleagues and contacts from other meat companies, who guided me through the meat company's software. We got what we could, which wasn't much. One snippet of information we stumbled across was a killing sheet that seemed to show that a very high-ranking New Zealand politician's farm was paid about 10 times the going rate for a line of lambs — one way of making covert political donations, we speculated.

Meanwhile, I had to front the media. This was all new to me, hosting TV news conferences and trying to explain to farmers that, while we were trying as hard as possible for them, the loss of their livestock and the proceeds was nothing to do with us. And I had to deal with the meat workers' union, which was understandably outraged on behalf of its members. A resource centre was set up in Hastings, and feelings ran high at the meetings that were held there. Each time I attended, I was acutely aware of how many big, strong men accustomed to hard physical yakka were there, and how angry they were. It was my job to assure them that Federated Farmers was doing its best to help.

Amongst the various claims the union made was that the laid-off staff from Tomoana were going to go hungry and would need urgent assistance, probably even food parcels. Because this was all unfolding around Labour Weekend, the traditional time to plant spuds in Hawke's Bay, I hatched a plan to use the holding yards at Tomoana, made fertile by the manure from generations of lambs, for planting a big crop of spuds to tide everyone over. I got permission from the receivers to use the yards, and I organised a supply of shovels and seed potatoes from one of the stock firms. But when the time came to put shovel to dirt, no one showed much interest. I was extremely disappointed, to say the least, but what made it far worse was that I received a barrage of phone calls from union officials accusing Federated Farmers (and me in particular) of indifference to the meat workers' plight. You grow a thick skin quickly in that situation, or you don't last.

As the catastrophe receded and came more into perspective, it could be seen as just further pain suffered by an economy that was struggling to adjust to turning out products that the market actually wanted. It was another death knell to the old model of New Zealand fat-lamb farming. The loss of Tomoana proved to be an amputation, not a *coup de grace* to the Hawke's Bay economy.

CHAPTER 11

MAKING MUDDY MONEY

In my early days in charge on Waipari, I was on the lookout for anything and everything that I could do that would add an income stream. It was probably a reflection of the entrepreneurialism instilled in me as a child, right from the days when I would trundle Celia's pram full of empties along in the Canterbury pre-dawn. It's hard to break the habit of a lifetime, and it's taken me in some strange directions.

An early and obvious way to raise a bit of grocery money was to let out one of the unused cottages on the station to families looking to stay on a farm to get some real Kiwi farming experience. In about 1987 we had a family in the cottage, and I happened to drop by to tell them that I was going to do a bit of maintenance on the limestone springs that supplied Waipari's water (and which, incidentally, gave the station its name: one of the meanings is 'water flowing from the hill').

'You might want to fill up a few bottles,' I told them. 'Things get a bit stirred up, and it could be pretty cloudy tomorrow.'

In the event, they didn't follow this advice until after I had performed the necessary maintenance on the springs — and then they complained that there was some sediment in the glass bottle they had just filled.

They were paying guests, and rather than follow my natural instincts to do a full Basil Fawlty on them, I joked that the minerals in the sediment were good for them, and that people paid bloody good money for water like that in Europe. Amongst many other health benefits, I said, it would give them strong teeth and nails!

'Oh,' they said, looking impressed. 'Do you mind if we fill a few more bottles to take back to Wellington with us?'

'Nah,' I cried expansively. 'Fill your boots!'

The lightbulb sort of went on. When I'd been in Europe on my OE, I had been bemused by the sight of fridges stocked with expensive bottled water. I couldn't believe people would pay for something that New Zealanders take for granted — or did in the 1980s, anyway.

And in saying that Waipari was blessed with the kind of water that got sold in expensive bottles in France, I was speaking true words in jest. At the top of the main range that runs through our station, Papahope, there is a series of naturally flowing limestone springs that are fed from beneath the other four aquifers in

Hawke's Bay, reportedly all the way from Lake Waikaremoana some 250 kilometres to the north.

There and then, I decided to investigate what would be involved in bottling and selling Waipari water.

I had the water analysed, and then had the chemical composition compared with those for certain other, overpriced, imported mineral waters on the market. I was most encouraged to find that Waipari spring water was equal to or better than the most popular French products in every respect that mattered.

Confident in the quality of the product, I did some preliminary market research (which was encouraging) and even went so far as to apply for (and obtain) a Business Development Board Grant to develop the project further. That paid for a trip to Auckland, where I met key people such as supermarket purchasing agents and Air New Zealand food and beverage managers. They, too, gave me cause for optimism. It seemed that the market for bottled water had turned against the foreign product in the wake of the *Rainbow Warrior* incident (where French agents had sunk a vessel belonging to the environmental organisation Greenpeace in Auckland Harbour on 10 July 1985), opening up an opportunity for local water.

There were very few others in the bottled water game in New Zealand back then. There were no industry guidelines whatsoever, so I simply compared notes with Richard McDonald, another entrepreneurial farmer who was going down a similar path using water from a Canterbury aquifer. I was also lucky enough to have assistance from a Hawke's Bay winemaker, Mission Estate's Warwick Orchiston, who gave me a few clues on how to build a bottling line that would maintain the quality of the water into the bottle. The options seemed to be to buy an expensive ($140,000) Italian bottling machine or settle for a manual system that was more labour-intensive but cost less. With interest rates still running at around 15 per cent (down from 18 per cent a few years before), the choice was simple. It also seemed as though it would provide a good, local employment opportunity for the wives of the station staff. So I roped in Keith Gosney of Harris Machinery in Hastings to help design and put together a sterile water-bottling plant, and meanwhile I had Waipukurau Construction erect a small factory building on Waipari.

When all was ready, I assembled the various bits and pieces of filters and sterilising kit that comprised the purpose-built bottling line and we did a few trial runs. There were the inevitable teething problems, but once they were sorted out we had a product we were proud of, and I went public with my new Waipari 'oil well' and sent our first consignment of bottled water to market, proudly wearing the name Mineralé.

Word got around the district, as it does, and I gathered I was a bit of a laughing stock there for a while, because the general opinion was that, as one highly experienced Hawke's Bay farmer told me one day with a pitying look in his eye, 'You can't sell water, Mark.'

I kept quiet, because I knew different. Several Auckland supermarkets were selling each of our 1.5-litre PET bottles for $2.30. Our cost of production was about $1.32, and allowing for the supermarket margin and GST, our water was fetching more than petrol and nearly as much as beer, and was free of excise. And it so happened at the time of the Auckland water crisis that I was standing next to that same farmer in the Waipukurau branch of the farm merchandise chain Farmlands when Chris phoned to say that Progressive Enterprises had just increased their monthly order from one pallet of Mineralé to 10.

I finished the phone call and turned to the doubter.

'O ye of little faith,' I crowed, or something like that.

We responded by triple-shifting the plant, and with Chris, Anne Allen (the wife of a tractor driver on the next station) and others working like Trojans, we managed to meet the hugely increased demand — until Alex Harvey Industries in Auckland announced they simply couldn't keep up with the increased national demand for PET bottles. That clipped our wings, but it all soon became academic, as one of our competitors, who claimed that their water was tankered to their Auckland plant from a spring in the Mamaku Ranges in the Bay of Plenty, was found to have resorted to diluting their product with Auckland tap water. The media coverage of that killed the market overnight. Some producers got caught with excess stock and others tried unsuccessfully to export it. We were repeatedly asked to reduce our price, to the point where I considered the price unrealistic. Rather than produce at a loss, I elected to mothball our plant. We were lucky that our overheads were low enough to allow us to do so. By the time the loss leaders had cleared their stock and demand was beginning to build again, I was too thin on the ground with other projects.

I could still smell money in Waipari water, and there was evidence that others could, too. One of the two goes we had at a joint venture — this one with Bronic Hasler, who already had a company distributing soft drinks and similar products around the North Island — looked promising. Bronic put a huge amount of effort into developing a new brand, PUREAZ, but not long after we brought it to market, Coca Cola Amatil New Zealand threatened him with papers indicating they intended to sue, on the grounds that we had copied their product Pump by selling a product in a blue bottle whose name also started with the letter 'P'! Bronic was a seasoned campaigner in this sort of thing, and when it became clear he wasn't taking it lying down, the lawyers' letters dried up. But by then it was clear that our product was being blocked by 'exclusive sales' agreements, whereby the big companies provided outlets with the chillers in which their product — and only their product — was displayed. We focused on the smaller outlets, but the costs made it uneconomic. When Bronic's delivery truck blew up, we canned the venture and mothballed the plant again.

We may even have attracted the interest of the French. One sunny afternoon, as I was droving a mob of lambs along the road, I came across a funny little man with a map spread out on the bonnet of a rental car. I stopped, as one does in the country, to offer assistance. In broken English with a distinct French accent, he explained that he was a rockhound, and asked how he could get to a certain limestone formation that was marked on his map.

I told him he might have a few problems getting his car up the track, and suggested he go to the other side of the district where the formed road finished closer to his destination, making for a shorter walk. It was only after he had bid me *'merci'* and *'au revoir'* that it occurred to me his 'interesting rocks' were precisely where our precious spring was. The *Rainbow Warrior* sprang immediately to mind, and so I hurried home and rang a few of those who lived on the road to which I had directed him and encouraged them to do him no favours whatsoever. Nothing blew up and no one died or got sick from drinking the water, so I suspect that the 5-kilometre scramble 1500 feet straight up in the heat of the day deterred him. I never heard anything more of him.

Another day, a very smartly dressed Maori gentleman knocked on the door of the homestead and introduced himself as from the Waikaremoana area. I invited him in and offered him a cup of tea, even though I was smelling, possibly, an opportunistic rat. Sure enough, as he waxed lyrical about his grand plans for securing the economic future of his iwi by extracting value from the whenua, he steered conversation delicately around to enquiring about how we were doing selling water from Lake Waikaremoana.

'You know,' I said, 'it's great timing that you dropped in today. You might be able to help with the project.'

He nodded encouragingly.

'You see,' I continued, 'I've just heard we might be going to be sued under the Resource Management Act for the damage our creek caused in the last flood to prime farmland downstream. If our water comes from your lake, then perhaps we can take joint financial responsibility.'

He stopped nodding, put down his cup and stood up.

'Nice to meet you, Mark,' he said, extending his hand.

I heard the wheels of his car spin in the gravel outside, such was his haste to be gone.

On yet another occasion, I was approached by a reputable Hawke's Bay person who was acting as an agent for a Chinese business contact who was keen to develop the New Zealand mineral water business. Could they visit? I was amenable, so the delegation of Chinese businessmen he was hosting added Waipari to their itinerary.

The group arrived on the appointed day in two big, black, shiny cars. (I half-expected them to have flags fluttering on the front quarters.) I was introduced to the leader, a very distinguished-looking gentleman named Mr Tong. I showed them over the bottling plant, and they seemed impressed. As long as we made a couple of modifications — an extra ventilation fan and a different doormat — they declared they would be ordering many containers' worth of business each month.

I suggested a trip up to the top of the hill in the Patrol to inspect the spring itself. They were keen at first, but as the track got progressively steeper they seemed tense. As I approached the really steep bit, they began making all kinds of waving and throat-cutting signals, which I interpreted to mean they were happy to skip the inspection of the source. I tried to turn it to my account by making a strategic diversion on the way down through a paddock of fine-looking, 18-month bulls, just in case they were in the market for some premium grass-fed beef. Some of the boys were conspicuously randy, and I did my best, in rather crude sign language, to imply that anyone lucky and wise enough to eat beef from these bulls might end up just as muscly and sexually active. But one bull seemed to think that we were more interesting than his mates and ambled over for a chat. I was well accustomed to such inquisitiveness, but my Chinese guests were convinced he was contemplating a murderous rampage. It was quite a sight to see all six of them trying to get into the Patrol all at once and through the same door!

Back at their motorcade, they produced white hankies and proceeded to wipe bull poo from their shiny patent-leather shoes as Mr Tong explained that provided we made the requested modifications, they would be pleased to accept one container of bottled water per month on a trial basis.

Our delight was short-lived. Just as we were about to fill the containers, the New Zealand agent rang to check on progress.

'Our buyer is looking forward to the arrival of the product,' he said. 'But just so you know, he has discovered he can get water of the same quality from another supplier for only 66 per cent of your price.'

I had been warned that this might happen.

'You tell your Mr Tong he can get stuffed,' I told the agent. 'We're not playing that game.'

'I can't say that!' was the scandalised reply.

'Well, I don't care how you tell him, but you let Mr Tong know that if he wants to play in our sandpit, he plays by our rules.'

Later, I heard that another company had got caught with their containers sitting on a foreign wharf, with the overseas agent claiming that there were inconsistencies with the product. Either they could drop the contracted price to 66 per cent of the agreed price or they could pay the cost to dump it. The whole episode was an insight into the difficulties the meat-exporting companies have when trading in foreign markets. I was inclined to praise the Lord for New Zealand's straightforward business practices!

※※※※※※※※

While I was doing market research for Mineralé water in Auckland, I was alerted to the fact that there seemed to be no New Zealand women's perfume that conjured up memories of Aotearoa for foreigners who had visited. According to many I spoke with in the high-end women's 'image enhancement retail business', there were a considerable group of customers who were relatively price-insensitive, who would be happy pay for a unique product if it helped make them stand out from an already exclusive crowd. Listening to the advice, I realised that many of the very successful and expensive women's perfumes came from sun-drenched areas that also had a vibrant and world-class wine industry.

A lightbulb moment again.

I had always enjoyed the heady scent of the cabbage tree in early November. Others agreed.

By adding in a few other flowers to the test bunches, I managed to find a blend that seemed to excite both male and female 'test drivers'.

Stage two was to find out how we could capture that smell in a packageable way.

Once again the Business Development Board showed interest and provided early support. With the help of some very enthusiastic, glamorous and 'price-insensitive' Remuera women, we continued the research, with my investigation leading me to an old retired French perfumer in Auckland.

The traditional way to make perfume, apparently, is to boil flowers in wax to extract the essential oils, and then distil these off for further blending. With this new information, I did costings and budgets, and found a potential marketing partner. It was to be a great new cottage industry with locals paid per kilogram to go out early morning in the springtime to harvest the various fresh native flowers. Even the name was reserved on the trademark register. 'Kaabargee', pronounced 'car bar-gy', sounded exotic and product-descriptive. It was exciting stuff to be able to harvest flowers that were underutilised, provide part-time employment for some locals, and produce a product with a retail value of hundreds of dollars per kilogram. Vague thoughts of a jet-setting lifestyle crossed my mind.

When discussing the project with Christchurch-based scented-soap manufacturers Glenys and Keith Wyatt, who had a shop selling their naturally scented wares in Merivale, they said to me that while the product was likely to meet very good demand, the base ingredient, in this case various native tree flowers, could be synthetically manufactured for a fraction of the cost of using the real ones.

My heart sank. Further investigations confirmed that information to be true, and I sadly closed the file, cancelled the rest of the project and placed the information gathered so far into the bottom draw of my rather untidy office desk.

Another project mothballed.

One day, at a district party, I was congratulating my friend Kate Worsnop on her acquisition of a nice, shiny-new Range Rover.

She made a face.

'Trouble is, Bruce won't teach me how to drive it off-road.'

'Oh, I can give you a few pointers,' I said. 'Bring the Rangey over next week and we'll have a play.'

Kate duly showed up with her vehicle a few days later, and we spent the afternoon sliding around on a muddy hillside while I showed her what the various levers and buttons did. As in all good test-drives, we did find the limits of its capabilities: I had to walk a few kilometres home to fetch the tractor to do a subtle recovery. But she finished with a new-found respect for what her Rangey could do. Afterwards, we went back to the homestead so that I could give her a brandy to settle her nerves.

'You should charge for lessons,' she said, as she gave me a very nice bottle of Te Mata red for my troubles.

'Nah,' I replied. 'You're a mate. I was happy to help.'

But Kate's comment sowed the seed. A few months later, I was watching staff from the Hawke's Bay Regional Council Taradale Depot carting 1080 poison up the hill using just a small motorbike while a perfectly good Land Rover sat parked at the bottom of the hill. I rang their supervisor, Keith.

'Why aren't you using the Landy you've got there?' I asked. 'It'd be much more bloody efficient than that stupid little bike.'

There was an awkward pause.

'We're a bit scared of the Rover, to be honest. None of us know how to drive it off-road properly.'

'Oh, I can give them a few pointers,' I said.

'Great! Put something down on paper and send it through. We might even be able to pay you for it.'

So that's what I did. I agonised over how much to charge. The usual farmer's rate of recompense is a crate or two of beer: as a species, we're notoriously generous with our time. But it so happened that I had had a bit of a tussle with the chairman of the Regional Council, who had got wind of the fact I was generating revenue from our spring and was insisting I pay $500 for a resource consent and the water right to the 1000 litres per day we were bottling. As the water was simply disappearing back into the ground after bubbling up from the spring, I felt it wasn't a tributary to the river, and therefore felt that the law relating to water rights was being misapplied. Neither side had backed down.

So I decided to charge the council $510 for the proposed day mucking around in 4WDs, with a $10 discount if they paid promptly.

They paid immediately, and booked another few courses besides the one I was offering the Taradale depot, so in the end I pocketed $1500 over and above the sum I grudgingly handed over for my water right.

The training day was a hoot. I did this stuff for fun, so it was pretty gratifying to know I was being paid to indulge in what was essentially a hobby. The council staff enjoyed it, too, and it was just as gratifying to see their confidence and skill levels improving as the day wore on.

I didn't really think much of it, but word soon spread and I began fielding calls from private individuals and organisations whose employees were occasionally called upon to drive off-road — notably Fisheries officers, the local power companies and motor companies.

One day, I received a call from Greg Hunt, who said he was from the marketing department of Mazda New Zealand.

'We've just launched a new range of 4WD utes,' he said, 'but we don't have anyone who's got much experience in four-wheel-driving. Could you help?'

'I reckon,' I said. 'How did you get my number?'

'You were recommended.'

He named the editor of a new 4WD magazine, who had just been on one of my courses.

'He spoke very highly of the course and of you.'

'I'm flattered,' I replied. 'But you should know I'm already talking to Nissan. I drive a Patrol.'

'I don't think that matters,' Greg said. 'If we send you down one of our new utes, can you evaluate it for us and report back?'

Why, yes. Yes, I could.

According to the arrangement we made, I picked up the nice, shiny Mazda Bounty SDX 4WD fresh off the transporter in Hastings. That night, I was due to attend a big Wool Board function being held at the convention centre on the foreshore in Napier. I was increasingly impressed with the Mazda as I drove the 20 kilometres there, and it got me thinking. Land Rover was one of the evening's event sponsors and had arranged for a shiny new Range Rover and Land Rover to be parked at the entrance to the venue. What if I parked the Mazda at the head of the beach on the other side of the building so that anyone wandering out to get a breath of fresh air and a glimpse of the sea would see it? I decided to satisfy my curiosity.

The beach area in front of the convention centre was well protected with high walls, but I was getting increasingly keen on my notion. I found an access to the beach a few kilometres away and proceeded to coax the new ute to chew its way through the very soft beach shingle along to the area of darkened beach right in front of the main conference room. I parked the vehicle at an angle that best presented its lines, locked the door and wandered, whistling, up to the venue to join the fun.

Little did I know that the order of ceremonies called for the conference room curtains to be pulled back and the floodlights turned on to illuminate the waves breaking on the stones 50 metres away. The shiny new double-cab Mazda Bounty ute was hard to ignore, framed squarely by the windows and sparkling in the lights.

'Oops,' I murmured, and found a dark spot in the room where I could skulk as people wondered how the hell the vehicle had got there. The Land Rover people were wondering the loudest, and didn't seem best pleased. But I consoled myself with the thought that they could have pulled off the same stunt themselves if they'd had the imagination and/or the ability to do it.

I waited some time for the majority of the guests to drift away before sneaking out, climbing in and retracing my tracks down the beach. It had never occurred to me that I would get stuck on the way up, but I was terrified that I might on the return journey, whereupon I would end up with egg on my face and most probably attract the attention of the law. But I got off the beach and drove home, quietly pleased with the marketing coup I'd pulled off.

Mazda were pretty happy with me, too, and they appreciated the thought and attention to detail I put into my evaluation of their product. The company top brass paid me a call, and indicated they would like to have me in the marketing team. I'll never forget the impatient wait I had standing beside the station's fax machine as the antiquated phone lines struggled to deliver the contract document. It scrolled out in a painfully slow coil, as I craned my neck to try to read their offer to become the 4WD adviser/consultant assisting the company in their effort to establish their product in New Zealand. I had always taken a pretty dim view of 'consultants', but we were just then clawing our way out of the very harsh drought of 1995. I had decided to buy another parcel of shares from the partnership in Waipari, and to say finances were tight is an understatement. As is often the case, the interest bill was greater than the income from the farm, so the opportunity to generate some kind of off-farm income to pay the groceries bill was very welcome. I figured if I couldn't beat them, I might as well join them and become a consultant, too. I formed a company called Hillseekers 4WD NZ Ltd to be the operations entity for the 4WD training and advisory side of things. In return for my services, I was paid a generous monthly retainer and a fee for each day I ran a training course at one of Mazda's nationwide network of dealerships. I was also given a ute to use as though it was my own. It was nothing short of a godsend.

I got on well with the Mazda marketing department's Greg Hunt, and together we devised the Mazda 4WD training programme, which entailed setting up three of their new 4WD utes with the correct off-road tyres, bull-, roll- and rear-bars and having the "Mazda 4WD training" name and logos painted on them. We also put together a very professionally produced instruction video, *The Steep, Deep, Rough, and Slippery*, and I compiled a Mazda-flavoured book of 4WD driving techniques.

Our first training event was held in the bush at Ardmore, south of Auckland, where we set up a camp in a remote clearing, complete with a huge, fancy, architecturally designed marquee (with a very pointy summit that we nicknamed "Madonna', due to the close resemblance it bore to half of the bodice the rock star wore on stage) with a large barbeque and generator-powered fridge for the refreshments.

Ten assorted Mazda dealers and clients would come in in the morning. We'd start with a briefing. You'd usually have a fair few in each group who fancied their own untutored off-road skills, and they would stand there with their arms folded and a cynical smirk on their dials as Greg and I delivered the briefing. But the centrepiece, of course, was the driving. I would take the cockiest and most sceptical attendees, leaving the rest to be divvied up between Greg and another carefully selected driver. We'd load up, four passengers to a vehicle, make sure everyone had their seatbelts on nice and tight, and then set off.

I'd make casual conversation as we bumped and lurched along the track, and then, without warning or any particular reason to stop chatting, the ground would vanish in front of the vehicle as it plunged over a lip and onto a slippery, steep (45-degree) drop-off.

You don't often hear grown men scream. I always enjoyed the expressions on the faces I could see in my rear-view mirror at this point! The language wasn't printable.

After this baptism by fire (or, more accurately, mud and gravity), we'd drive them over a series of carefully selected, surveyed, pegged and tested obstacles, including a steep (40-degree) hill-climb and a sidle across a 35-degree hillside.

By now, we'd have their attention. Even the most been-there, done-that types would be putty in our hands.

Next, we'd take them to the second, more sedate course, where the safety margin was considerably greater. Here I'd show them the wrong way and the right way to do a stall recovery and a slide. I'd explain how to drive out of a roll by steering the nose of the vehicle quickly downhill if the vehicle became unstable, and how to regain control as the ute slipped way, way down a steep, slippery slope by accelerating quickly, keeping the wheels turning at ground speed.

Then it was their turn in the driver's seat. Each driver would put the ute over the same set of obstacles, and would receive individual instruction on handling the vehicle in terrain they would never previously have thought possible to drive. After a few hours of vigorous and sometimes apparently terrifying off-road adventures, we would deliver the somewhat more subdued and hungry passengers and pupils back to the bush camp for a fancy barbequed lunch just as another 10 were turning up for their afternoon course. It was always fun watching the new intake listening sceptically to the stories of 'how the ute disappeared over a cliff', 'how I managed to drive out of a roll-over', 'how I climbed a bloody cliff in a ute', with the cliffs getting higher and steeper and the threat of rolling more imminent as lunch went on.

After an hour of this softening up, it would be the turn of the afternoon group. You could sense the apprehension before you hit the drop-off. (In one case I could even smell it, as a vaguely familiar but definitely unmechanical smell suffused the cabin. It became plain who was suffering as soon as the view changed from bush to sky and the ute lurched into a nosedive. One of the passengers — a female, as it happened — unleashed the loudest, most blood-curdling scream I have ever heard.)

At the end of each course, our students sat a 25-question, multiple-choice test and were issued with a copy of the 4WD book. If they passed the test, they were also later issued with a Hillseekers Mazda 4WD Training Programme certificate.

After we'd been running the courses at Ardmore for a while, we took the concept on national tour. We loaded the vehicles onto a transporter and shipped them to Invercargill, with Greg and me following by air. Mazda had once enjoyed legendary status amongst Southland farmers, but this had fallen away. The hope was that we could restore the brand to its former glory. After running a few very well-received courses in Southland, we made our way up the island. I saw parts of the country I would never have seen otherwise. I enjoyed the novelty of living in motels and eating at fancy restaurants at Mazda's expense — for a while. But it wasn't long before I began to notice the difference in your waistband that comes from taking a break from the usual round of hard, physical exercise and sitting on your bottom being tempted by restaurant meals.

Still, the programme was a great success. It did the heart good to see the car park filled with vehicles manufactured by the competition, and dealers reported being inundated with calls from people wanting to enrol. Bums on seats was the name of the game, and some dyed-in-the-wool drivers of the opposition's products found themselves surprised converts to the Mazda badge by the end of a day out with us.

The Toyota Hilux was king when we started out, and was being promoted with a very popular series of TV advertisements featuring legendary Kiwi bushman and raconteur Barry Crump at the wheel and a whole lot of trick photography to make it look as though he was performing unbelievable feats in his Hilux. It was our instinct that proving the Bounty in real-life conditions would serve us better, but that wasn't to say Mazda was going to neglect TV advertising. In fact, a whole campaign was devised around the lines: 'The new Mazda Bounty ute. It's enough to make Crumpy grumpy.' But Crump died just as the ads were to go to air and, out of respect, Mazda pulled the campaign and all the printed material was dumped. I later heard on the grapevine that Toyota thought our product wasn't good enough to survive in that market, anyway.

If they ever uttered them, Toyota was made to eat those particular words a few short years later. It was a proud moment for me when I learned that combined sales of the Mazda Bounty and Ford Courier — to all intents and purposes a Mazda vehicle badged as a Ford — had exceeded Toyota's.

Taking the Mazda course opened up other opportunities, too. Just before we were due to run one at Ohai in Southland, Greg indicated that I should take particular care of one of the people who was booked to attend. Phil Dunstan's Lakes Contract Services had just won the contract to operate all of the Queenstown/Southern Lakes district council services, including snow-ploughing.

After an afternoon of showing him off-road driving techniques, Phil asked, 'Do you reckon you could train up my snow-plough drivers?'

I shrugged. All those years I had misspent flogging more or less unsuitable vehicles up and down skifield roads had given me a pretty firm grasp of the principles of extreme winter driving.

I figured I could probably modify the 4WD course to offer a version for drivers who regularly faced heavy snow and ice conditions, such as first-responders.

'Yeah, I reckon I could.'

Two years later, about 50 people enrolled in the first of what I came to call 'Freezedrive' courses. The first component was a theory session held in the Lake Hayes hall, where my second wife, Sue, who had a background as a professional ski patroller and avalanche forecaster, explained the mechanical properties and types of snow. I followed with a PowerPoint presentation on the cause and effect of slippery conditions as they related to vehicle control. I allowed plenty of time for those who attended to ask about the features of the different types of 4WD vehicles on the market, their advantages and disadvantages, and why results so often varied from those promised in the glossy sales brochures!

As happened in all our courses, there was a group of hardened old drivers, all with extensive experience gleaned from a lifetime of driving in just such conditions, who clearly didn't expect to get anything more out of the course than the sausage rolls we served at morning-tea time. They stood at the back of the hall with their arms folded, doubtless wondering why on the earth they were being made to listen to a North Islander telling them their job. But I noticed as one by one they unfolded their arms and leaned forward as their cheekier colleagues tried to trip me up with technical questions. My answers seemed to impress them.

We ran two practical sessions a day, one in the morning, one in the afternoon, with 10 drivers on each. They would turn up to the Remarkables skifield car park in their snow ploughs — seeing all those big machines parked there was a dream come true for a man who had longed to handle such machines when he was a boy playing in his sandpit. I would put the drivers through their paces on a sheet-ice skid-pan I had constructed in the car park, and on a course which required them to drive in deep snow on steep tracks. I must have won over the doubters, because one day as I was driving to work up the Remarkables Road I was half-listening to the local radio station on which the Lakes Council Services Operations Manager, Garry Healy — otherwise known as Captain Gravel — was giving his usual witty weather and road conditions report, when he happened to mention that all of his drivers had been on a winter driving course with some stranger, and that they had been surprised to learn lots of useful new tricks. I tuned in. I was wondering how I could get in touch with this knowledgeable stranger when it suddenly dawned on me that it was me he was talking about!

One morning, when we were getting ready to leave the LCS depot at Frankton to head to the skifield, Phil suggested we swap vehicles. He would take my heavily modified Nissan Patrol for a spin to see how it handled on the Remarkables Road. I would get to drive his ride — an American Ford F250 Super Duty pick-up truck fitted with a bright orange, six-way, multi-angle Western snow-plough blade mounted on the front. I was delighted to accept. Phil took off from the Frankton yard, leaving me to familiarise myself with all the electronic controls of this super-sized Tonka toy as I went. I hadn't quite figured out how to feather the blade by the time I reached the historic Kawarau

River bridge, and as I lined up on the single lane with the 9-foot-wide blade fully presented, I was acutely aware of how fine the margin for error was. With only a few millimetres clearance on either side, I didn't dare fiddle with the controls in case the blade slewed sideways and collected the wooden parapets. So I concentrated fiercely on the line I had to steer, barely noticing cars approaching from the other direction beating a hasty retreat when they saw the monster truck with its car-eating blade sweeping towards them from the other end of the bridge.

Once across, I realised I had been holding my breath.

As soon as it was safe to stop, I pulled over and worked out how to fold the blade into a neat V before carrying on.

When I reached the top of the Remarkables skifield road, Phil asked me why I had taken so long to climb the hill. I told him I had to stop after the bridge to work out how to fold up the blade.

He looked a bit incredulous that I'd driven across with the blade fully extended.

'We didn't think it could be done!' he said.

That explained the looks on the faces of the car drivers!

***

It has often been said that the sound of snow falling is eerie. After I had been in Queenstown for around a week — a hectic week it had been, too — I woke at five one morning to that distinctive, muffled hush. I peered out the window of my hotel room with its view of the Steamer Wharf and the lights of Queenstown's central business district and found the prospect softened by fat, drifting snowflakes. As I arranged my bed so that I could watch as Mother Nature tried to erase the world, I became aware of flashing orange lights glancing from afar and getting ever brighter. Soon Phil's illuminated mechanical monster with its hungry plough blade came roaring past, parting the building snow drifts with ease, leaving a trail of divided snow banks behind it. The flashing warning lights and glaring floodlights made for an impressive and intimidating sight. It was inspiring. I sprang out of bed, had a quick, hot shower and then flung open the doors and stood there on the balcony as snowflakes settled on my still-steaming, naked body. It's not too often you get to do that in downtown Queenstown! My exhibitionist reverie was rudely interrupted by the ringing of my phone. It was Phil.

'What're you up to?' he asked, doubtless expecting me to be still in bed.

'Oh, I'm standing nude on my balcony watching you plough the snowy highway,' I answered. 'I just saw you drive past in your Tonka toy!'

'Hurry up and get dressed,' he said. 'I'll pick you up in 20 minutes and we'll do a tour of duty in the plough.'

As I flung on some clothes, the fire sirens started sounding all over Queenstown, to warn of a major emergency afoot. The phone went again.

'Forget the fancy fun stuff,' Phil said, his voice indicating he meant business. 'Get up to Fern Hill ASAP. We've got real carnage up here.'

The phone then went dead.

It took only a few minutes to drive the Patrol south along the lake front to Fern Hill Road. Around halfway up, on the steepest part, a full-scale vehicular smother was in progress. Drivers, unaware of how slippery the snow was making the road, were hitting the steepest part of the descent without realising that their brakes were next to useless. They were breaking traction and sliding helplessly into a mounting pile of vehicles at the bottom of the hill. One vehicle had gone over a steep, bushy bank, and the passenger was still trapped inside. As I came up the hill, I found Phil desperately trying to get people to stop before it was too late. He handed control of the situation to me and carried on up the hill to spread grit and to try to close the road at the top to put a stop to the increasing carnage. As most of the drivers didn't seem to have the skills to drive in control down the hill, I ended up bringing many of the vehicles down the road and running back up the hill between each drive, in order to clear the mess. I finished up completely out of puff. Unlike that time, many years ago, when I had dealt with a real sheep smother, this time I was out of breath from being overfed and under-exercised.

By now, the emergency services had arrived and were dealing with the aftermath of the main accidents. The worst was soon over, and to everyone's relief no one was hurt beyond a few bumps and bruises and lavish dents to both vehicles and pride. The pile-up was methodically picked apart, and as I looked on I was already formulating ways in which to use this real-life example in the Freezedrive course, so that people had the tools to avoid similar winter driving cock-ups.

One of the trainees on Freezedrive that week was Phil Jones, who was in charge of the Queenstown Southern Lakes police district. At the debrief at the end of his day out with us, he asked if we could offer an extended version of the course to local police, ambulance and fire rescue staff. That hadn't been part of my original plan, but putting my head together with Simon Richards, a police driving instructor, we developed a course for police from the Otago/Southland area.

As a result I found myself demonstrating how to control a Highway Patrol Commodore performing perfectly balanced figure eights on sheet ice, sometimes with the speedo showing poacher rather than gamekeeper numbers. I let Simon do some of the hard yards riding shotgun in the hard-sprung 4WDs. He was a well-built, fit sort of bloke, who rode motorbikes for fun, but he found that the demands of bracing in a lurching, slipping, sliding 4WD, all while craning your neck to see what the trainee was doing with their feet on the pedals, were taking their toll. He told me he was taking Voltaren by the handful at the end of each session.

One day, I had just stopped to take aboard another police officer when my cellphone went. I usually ignore calls when I'm busy, but this was from College. That struck me as unusual, so I took the call. It was Simon Leese, the headmaster.

'I've rung to offer your son a scholarship,' he told me, after the usual pleasantries.

Just then, one of the cops playing around turned on the siren of a car behind us to announce that they had completed the challenge without coming to grief. I broke off the call briefly to welcome my next trainee aboard.

'Jump in, mate. Strap the belt on tight. You'll be right,' I said, then re-focused on Simon.

'I'm hardly going to say no, am I?' I said to him. Things were very tight on the farm, and enrolling my eldest son William at College for what I considered to be the best possible education was a huge financial leap of faith, as we would still have to pay a proportion of the fees.

There was a pause.

'Have I rung at an awkward time?' he asked. 'Sounds like you've got some sort of emergency there.'

'Nah,' I said. 'I'm just training the police.'

He later commented that whenever he phoned, I always seemed to be in the middle of some drama or another.

After I had finished the call, I admit I had a brief, teary moment as the great news sank in. I apologised to my trainee — a young woman — for answering the phone, but explained what it was about.

'It'll make all the difference for Wills,' I said, half to myself.

'I know what you mean,' she said. 'My parents had to take me out of private school about halfway through the tough eighties, and we've always felt very regretful about it.'

These courses were enormous fun, but there was a serious purpose, too. We must have made a difference and prevented accidents and vehicle damage, as the Police Training Co-ordinator, Ingrid Adamczyk, kept on prising funding from the tight fist of the national organisation for some years to come. And one night when I was in a bar in Queenstown — off-duty, of course! — an ambulance driver came up to me, shook my hand and told me that a technique I had taught her had saved her, the vehicle and its patient from a potentially catastrophic spill off the side of an icy road. Similarly, an email I got from Andrew Burn, head of Highway Patrol for the Southern Region, told me that he had just about got stuck in the snow at the end of a long, cold shift when he had remembered a little trick I had told him. When you're stuck in snow, if you open your door and watch the ground closely with the handbrake pulled on a few notches — to both lock the differential and soak up excess power — you can apply just enough power to get the vehicle moving very, very gently without wheel spin. This made the difference between spending a bitter night shivering in his car and getting home to a shower and a warm bed. Feedback like that made it all worthwhile!

Meanwhile, too, word was getting out about the Freezedrive course, and it was heavily oversubscribed. Because the Remarkables skifield was available for only a short period of time every year, the window in which we could operate was tight. But in the time we were open, we had through the course all sorts of interesting types — including Boeing 777 pilot trainers and international ski-racing instructors, to name but two groups. And so many people from Queenstown attended that it wasn't too long before I felt I could wander down the street there and be greeted almost as a local rather than just another tourist.

I was also called upon to offer the Freezedrive course further afield. Late one wintry afternoon, as I watched the staff of the Mount Hutt skifield whom I had just trained start to edge their way down the treacherous, frozen mountain road in their

Land Cruisers, the sound of the tyre chains clanking in unison carrying across to me, I cast an eye out across the Canterbury Plains and to where I could see my old boarding school, Waihi, some 80 kilometres to the south-west. I remembered how badly I had longed as a 12-year-old to be able to afford to go skiing whenever I wanted. It vaguely occurred to me that the fee I had just received for one day's teaching Freezedrive was more than enough to ski for the whole season.

CHAPTER 12

BELIEVING YOUR OWN ADVERTISING

If you're ever short of a bit of entertainment, Google yourself and your business and see what the world thinks you've been up to. I was indulging in this kind of egocentric online curiosity one day and was taken aback to discover that Mazda NZ, along with Hillseekers 4WD Ltd as co-respondent, was the subject of legal action in an Advertising Complaints Authority case.

It came back to me in flash that Andrew 'Muddy' Clearwater (Mazda NZ's boss) had asked me to write a very formal letter to Mazda explaining my background, ability and experience and the preparations I had done for the filming of a 'Zoom Zoom' TV commercial for a 4WD ute the year before. I was guessing this was why.

And so it proved. Apparently some well-meaning but overly enthusiastic lawyer, who also happened to be the president of a very southern Land Rover owners' club, had seen the ad and decided that the driver was exposing the vehicle to a high possibility of rolling over, due to the line he had chosen to drive the vehicle across the hill.

Technically, he had a point, and I ought to know — I was the driver.

Mazda had rung up and asked me to put aside a few days to come to Auckland and film an ad for the new model 4WD ute. The advertising agency brief was that I would be dressed up as some sort of Bob the Builder character who escapes the drudgery of working in an East Tamaki boat-building yard to go to the rugged west coast for a spot of evening fishing off the rocks.

At six one sunny spring morning I was picked up from my hotel by the film director's assistant. She was nervous: it turned out it was because she was at the wheel with some sort of professional driving instructor (me) sitting there next to her in judgement of her skills as she negotiated the city traffic.

'I'm hopeless at sharing the road with anyone else,' I said, in an effort to put her at ease. 'You're handling it all better than I would.'

She must have taken me at my word, because she promptly organised a stunt double to do the city driving part of the ad for me!

We arrived at the film set at the boat yard. It was exciting, particularly as they had a very impressive Flying Trestle food truck set up dispensing every and any type of food I could wish for. There were about 17 people running around, looking very busy. Any thoughts that there might be a film star's trailer to retire to while I prepared were dashed, however, when the wardrobe assistant handed me a set of appropriate 'builders' clothes' to change into — cargo pants, a polo shirt and sneakers — and indicated that I should do it in an office. Once changed, I waited for my involvement to start, and amused myself watching, fascinated, as all the gaffers hurried around, each in their identical black grunge outfits, with every possible tool hanging off their skinny hips.

Then it was my turn. The first scene in the ad had me strolling purposefully to the ute and climbing in like an experienced builder, then starting off and hightailing it out of the yard at the end of the day's work. That was all easy enough. Only two shots were necessary to get a wrap.

The stunt double did the next part — driving the ute through Auckland traffic and out along the western motorway while being filmed from a helicopter. I wasn't offended at all!

The next location was a patch up in the bush where some rough tracks climbed the hill. The director indicated I should drive up them, and I did. Then the director indicated I should make it look harder and more brutal, so I did. Even that wasn't enough, so I did it again, this time somewhat recklessly. That got the thumbs-up.

Next came the money shot.

The brief called for me to drive as fast as I could down some 100-metre-high sandhills and out along a sandy creek through the bush and onto the beach. There, I would park beside a rocky reef, take my surfcasting rod from the tray and cast a line into the evening sunset.

I eyed up the sandhill. It looked a bit tame.

'What say I drive across it on an angle to make it look like the ute's about to roll over?' I suggested. 'I can drive it out of the roll, down the steep bit and into the creek. That'd look much more exciting.'

The director's eyes lit up, and he swapped glances with his crew, who were all nodding enthusiastically.

'Could you?'

'No problem,' I shrugged.

While they began discussing where to site the cameras, I carefully placed a couple sandbags into the tray of the ute diagonally opposite the front, downhill wheel to help it handle to my satisfaction. I backed up to the base of the sandbank to check for bumper clearance on the final exit. Then I eyeballed my route. There was a big patch of rocks just at the critical point where the gradient changed to close to 40 per cent with only one possible exit line passing through it. There was no margin for error.

A lunch break was called, and we were treated to a very urban picnic.

I outlined the upgraded plan and there was much oooh-ing and ahhh-ing from the advertising people as they contemplated the extra wow factor we were scripting in.

After lunch, I was handed two radio telephones: one tuned to the helicopter pilot and airborne film crew, and the other to the director and his crew on the ground.

At the top of the hill, ready to go, I surveyed my preferred option and a potential Plan B. The exit line was invisible from here. But just as I was about to walk out onto the pristine face of the sand dune to drop some cabbage tree leaves to serve as an unobtrusive visual reference, the RT crackled into life.

'Don't walk out there,' the director said firmly. 'No footprints allowed.'

Soon I was going to be careering over the brow of the hill effectively onto a sandy cliff. I didn't want to do it completely blind. The best I could do was throw some leaves onto the approximate line from a distance, and line up some trees on the far side of the valley, as a secondary, back-up focal point.

After double-checking that I had tied the extra weight down properly, I did a dummy run on a similar hill nearby to reassure myself that I had reset the tyre pressures appropriately for the job. Then there was nothing for it. I studied the line one last time to get it sorted in my head, rolled down the windows to avoid the tedious business of picking broken glass out of my clothes and hair if it rolled, and pulled my seatbelt as tight as possible.

'Ready,' I said over the RT.

'Stand by,' the chopper pilot came back. 'I just want to do another run through the line so we can check the camera angles.'

I loosened my seatbelt and breathed deeply, focusing on the job ahead. I knew we'd only have one chance to get it right, because after I'd done my bit, it would look like a very driven-on hill.

Four minutes went by.

'OK, we're in position,' the chopper pilot announced.

I retightened my belt.

'Roll cameras,' the director replied. 'And ... action!'

I took a very deep breath and launched the ute out across the sandy cliff.

As soon as I crested the easy part of the hill and hunted for my cabbage tree leaves, I knew that something hadn't gone as I'd planned it. In the same split-second that I saw this, I realised that the rotor wash from the helicopter had blown the markers away. I was blind after all, but there was no realistic chance of aborting. I was committed, and the laws of gravity were fully in charge.

I felt the ute getting very light on the top side, with the drive from the uphill wheels taking all of the power and offering no traction in return. Instinctively, I stabbed down on the brake pedal with my left foot to balance the traction and dug the spurs in hard with my right — an off-road rallying trick to maintain control — all while at the same time searching for the line of trees over on the far side of the valley that would serve as a clue to the safe line to take.

Through my open widows, I imagined I heard some of the agency people, looking on from around 75 metres' distance, start to scream as the ute gave every indication of entering a balletic barrel roll. But luckily I managed to snatch enough power and traction to bring the nose around downhill.

Roll carefully managed. Now all I needed was to be heading towards the gap in the rocks.

More by luck than by good management, and perhaps with just a fraction of a second's worth of quality control, I managed to impose my will on the downward plunge of the ute enough that it lined up with the gap. Gravity took over at that point, and the rocks whipped by in a blur on either side.

The bit that I had got wrong was the exit. Perhaps I was too used to the suspension in my own Mazda ute, which I had upgraded significantly. Perhaps I was just going faster than I had reckoned. Whatever the explanation, there was a bone-shaking crunch as the vehicle bottomed out at the point where the slope met the flat.

That can't be good for the radiator, I thought.

I was conscious that the chopper was still hovering only about 10 metres above me. The show had to go on.

The plan called for me to race down the sandy-bottomed stream bed, sending showers of water skyward to catch the late afternoon sun. I hardly paid any attention to my performance or to the rattle of the loose front bumper, so intently was I watching the needle of the temperature gauge creeping into the red. As I had suspected, the radiator was terminally damaged.

'CUT!' called the director.

'Thank Christ for that!' I radioed back. 'I've got a bit of damage. Temp's up, and I think the bumper's a bit loose.'

'Nah,' came the voice of a gaffer, whose job it was to guard a ford we had crossed to prevent other traffic getting in our way. 'The bumper fell off totally three crossings ago.'

Nursing the badly damaged, overheating ute back to base, I was a bit worried what the reception might be.

The director raced towards me as I pulled up, steam rising from the nose of the vehicle, and yanked on the handbrake.

'Abso-lute-ly brilliant!' he enthused. 'We all thought you were totally doomed when it started to roll. Some of the girls screamed!'

'Yes,' I replied. 'It was a bit different to how I had planned it. The chopper blew away my markers! Sorry about the ute. Bit embarrassing ...'

'Nah,' the director waved his hand dismissively. 'Don't worry about that! We'll just swap the plates from the hero ute to the one we filmed on the road and re-do the creek shots.'

So that's what we did. This time, I was able to concentrate on getting maximum splash effect as I careered along the stream bed — a perfect job to ask of an overgrown boy who had begun his apprenticeship for the role back in his sandpit. I managed to raise great, glittering walls of water on every bend.

The exit onto the beach required a good deal of planning, not least because the activity of the helicopter had attracted the attention of a sizeable crowd of onlookers. We were wondering how we were going to clear the area so I could go full bogan in the soft sand as per the brief when the chopper alighted and the director handed four big bags of Mackintosh's toffees to the cameraman. While the chopper delivered a massive, aerial lolly scramble at the southern end of the beach, we were left in peace at the northern end to get on with it.

Very clever, I thought. These film types think of everything.

The next scene required me to drive the ute along the last, bush-lined stretch of river bed and out onto the beach, where

I would roar to the base of a rocky spit about a kilometre south-west, pulling a long, perfectly controlled 90-degree power slide out of the creek and onto the beach as I went.

No problem.

I did a quick check to establish where the hard sand met the soft. This change in underfoot can catch the unwary and cause an unexpected roll-over and, besides, I needed a consistent surface if I was to achieve the X factor in the slide.

That done, I backed a fair ways up the river to get a decent run-up and waited for the director's call over the RT.

'In position,' the chopper called.

'Roll cameras,' the director answered.

'Rolling.'

'Action!'

I called upon every bit of intercooled turbo power at my disposal and launched the ute, positioning it in the shallows at the edge of the water to get maximum spray. Then, as the beach appeared in the gap in the scrub, I peeled off into the 300-metre, finely balanced drift, emerging from it to aim down the hard sand along the centreline of the beach to the reef.

'Cut! Cut!' came the call.

'What's up?' I asked, much surprised. The run sheet had called for me to 'exit onto the beach, and proceed to the reef as fast as possible'. I thought I had followed that to the letter.

'What speed were you doing?' asked the helicopter pilot.

I hesitated. I didn't know who else might be listening to the RT, and to avoid providing anyone with rock-hard evidence of

229

vehicular impropriety, I replied: 'Well, the number had a one and a four in it.'

It turned out I had been too literal. The helicopter pilot had to fly out of balance to give the cameraman the best angle, and in that attitude, the helicopter couldn't keep up.

'The cameraman says it's a pity,' the pilot reported. 'It would have been brilliant action footage!'

Time was now against us. Just as the sun was setting, I made my final run up the beach, keeping it closer to the open-road speed limit. As the cameras rolled, I whipped out my rod and cast a sinker far out into the waves. There was no chance of catching anything: there was no hook on the line, just the weight. The RT came alive with the words so beloved of everyone involved in the film and TV industry.

'OK, everyone. That's a wrap.'

There was no wrap party, because half of the crew were still wet from helping me repair the broken ute in the creek bed. Instead, we made our way back to the hotel. On the way, the driver told me that she had worked on set on the *Lord of the Rings* trilogy.

'This was much more exciting!' she said.

I was eventually called upon to respond to the ACA complaint. Yes, I wrote, the complainant was technically correct. At that angle the vehicle could, and was in fact starting to, roll. But I felt I had made enough provision for the outcome, and had enough experience to handle the situation. And I also invited the complainant to consult his club rally records, where he ought to find my name against two class wins and a second overall in their club 4WD rally in 1980/81.

I heard no more. But while I think many of us at Mazda were proud of the fact that we didn't resort to various forms of trick photography (as commonly used elsewhere in the industry) to show the product's mettle, Mazda didn't do any further filming in New Zealand. Perhaps it was a cost-cutting thing or perhaps they were scared off, but they settled for much more sedate — not to say boring — Australian ads, accompanied by lots of cautionary text.

That wasn't the end of all the fun times, though. One day, I was rung up and asked if I could squeeze in a few days' vehicle testing before Christmas, with Kirk Kawaguchi (head of the Mazda ute division of Japan) and his technical design team tagging along. I was told in no uncertain terms that it was a great honour to be included in the design process. Normally the design team didn't venture far from Japan. Coming to New Zealand was a big deal.

When I received more details, I learned that part of the reason they were going to these unusual lengths was that the opposition product was rated to tow a trailer weighing 50 kilograms more than the Mazda was allowed to tow. Mazda Japan were accordingly seeking some practical experience of what could be safely towed in real-life conditions, so that they could claim a higher safety rating for their product sold in the South Pacific. I was able to tell them that I often towed my jetboat on an unbraked trailer without any problems: at 1250 kilograms, this was almost twice the stated safe towing weight for an unbraked trailer.

For the purposes of the formal testing, I was instructed to gather three trailers, a couple with brakes, one without, and have them weighed at an official weigh station, variously loaded up with certain weights. Both measurements and records were to be very precise. Then, on the day of testing, three officials from Mazda Japan and two from Mazda New Zealand arrived at Waipari just in time for lunch. As we did the traditional business card presentation, Roger Russell, a technical adviser for the Mazda New Zealand team, quietly told me that one of the Japanese delegation team was the man who had developed the Mazda rotary engine, and who had personally built the engine that powered a Mazda race car to victory at Le Mans. I was in the presence of engineering royalty!

The first job was to take them around the station in my Mazda ute, showing them what we use a 4WD for and what modifications we would like to the production design. At one point, we stopped at the top of our highest hill, Papahope. As is the New Zealand way, I casually pointed out the station boundaries.

'So much land!' Kirk exclaimed. 'You must be a very rich man! Why do you work for Mazda?'

I explained that I owed the bank a lot of money for the privilege of owning my land, and assured him that I worked for

Mazda because I enjoyed being involved in vehicle design and development.

With the Mazda New Zealand team following safely behind in another vehicle, I took the Japanese party in my own vehicle and found as steep a hill as I thought it could climb and embarked on an ascent. Near the top, I put my foot on the clutch, allowing the ute to career backwards, seemingly out of control. As we were sailing backwards at a high rate of knots, I asked Kirk (who was in the passenger's seat) to pull on the handbrake. The seatbelts were self-locking, and the handbrake handle was mounted under the dashboard. It was practically impossible for him to reach it. I brought the ute to a safe standstill and politely but firmly explained that's why we wanted the handbrake mounted in the middle so that a passenger could pull it on in a situation where the driver was, for whatever reason, unable to do so.

'Ahhh!' my passengers all exclaimed. There was a flurry of Japanese spoken, then Kirk told me in his very good English: 'Now we understand why you want the handbrake here.'

Mission accomplished!

Then we turned to the towing/braking trials. With my boat trailer hooked on behind, I demonstrated that even on a steep shingle road, I could stop the combination of ute and trailer in the required time and distance. Happy with my demonstration, Kirk asked if one of his team could take over the driver's seat and run his own tests.

No problem, I said.

I wasn't quite so sure as, safely buckled up in the back seat, I watched as the man drove the ute at close to the gazetted open-road speed towards a hard (45-degree) left-hand corner on a loose shingle road, something that no driver experienced in New Zealand conditions would do. There was a major drop-off on the outside of the corner, and halfway through the bend, he lost his nerve and braked heavily, which had the predictable effect of jack-knifing ute and trailer.

I didn't feel in any real danger, with my seatbelt on and the very strong, chassis-mounted steel roll-bar passing just behind my head, but I was a bit worried for my precious jetboat.

Luckily, the grader had been through not long before and there was a deep windrow on the margin of the road. That saved us to drive another day!

The Mazda team had a robust 3D computer matrix requiring all of the elements of the equation to be tested in order to be signed off. While for reasons of commercial sensitivity I have fudged the numbers, we had three trailer weights (say, 750 kilograms unbraked, 1250 unbraked, and 2000 braked) and we were obliged to drive at speeds of 50, 70 and 100 kilometres per hour on each of the three road-surface types we have in New Zealand: tarseal, shingle and grassy paddock tracks.

All was more or less well until the question arose as to whether it was safe to drive at 100 kilometres per hour with a 1250-kilogram unbraked trailer on grass. I repeatedly tried to explain that we wouldn't normally drive on grass at anything like 100 kilometres per hour with a heavy trailer, but they needed a tick or a cross in that box. My protests fell on deaf ears.

'OK,' I said, deciding to change tack. 'Hop in.'

We were performing the grass tests on an airstrip with the boat tied on behind the ute. With the designers aboard, strapped in and with pens poised above their test matrix, I willed the ute to get up to 100 kilometres per hour before we ran out of strip. It was important we did, because at the end of the strip where a topdressing plane would expect to be wheels-off, the ground fell away dangerously.

The needle nudged 50, then 60, as the ute bounced and slewed on the rough strip and the void beckoned. It was like playing chicken. Seventy ...

'OK! OK! OK! Stop! Stop!' Kirk yelled.

I pulled up hard.

As we sat there, there was a slight scratching sound as a shaking Japanese hand put a wavery tick in the necessary box.

'Thank you, Mark-san,' Kirk said politely. 'We have achieved our objective.'

On our way home, I took a shortcut through a paddock where there was a mob of yearling bulls. Sensing the need for a bit of entertainment to lighten the moment, I stopped, took the .22 rifle I kept behind the rear seat for shooting possums and rabbits, and invited Kirk to do some bull shooting. Perhaps our Kiwi humour is hard to get at times; Kirk took me totally seriously. His face lit up. He took the (boltless) rifle from me reverently and raised it, settling it with the butt resting on top of his shoulder, as you would a bazooka, rather than nestled against it as you do with a rifle. It hadn't occurred to me before then that it was quite likely that none of these men had ever handled a firearm before. I couldn't properly remedy that on the spot, as I didn't have the rifle bolt with me, but my suggestion that we do some possum and rabbit shooting instead that night was greeted with great enthusiasm.

I picked the crew up after dinner from where I had arranged for them to stay, at Neil and Judy McHardy's farm-stay, and we set off to share with our Japanese friends the Kiwi sport of spotlighting for possums. As luck would have it, though, no matter how hard we looked, we couldn't find a single sacrificial possum. God knows where they go when you really want them!

But under strict supervision, and allocating only one bullet at a time, I invited each of the team to fire at a log in a dam. Much excitement and hilarity ensued, and many photographs were taken. So much of what we take for granted is way out of the ordinary for people in other parts of the world. To cap off their experience, Neil and Judy sent Kirk and his team home with a sheepskin rug each. Sometime later, they got a letter from Kirk thanking them. It was, Kirk told them, his 'ideas rug' — the one he lay upon in his office while he sought inspiration for the design of the next model ute!

When the time came to release the new model Mazda ute, now called the BT50, I was invited to Auckland to be part of the big launch to the media and dealers. As part of the event, I used a 12-ton digger to throw up a set of humps and hollows in the grounds of the Villa Maria winery to use as a specialised 4WD course that would show off the new vehicle's capabilities.

Kirk was flying in especially for the big event, and I was called to Mazda head office to be part of the welcoming party. I was a bit late for the official welcome — as I had told the director's assistant on the day of filming the ad, driving in city traffic isn't my forte — but as I sneaked into the main office hoping my lateness hadn't been noticed by the distinguished guests, Kirk spotted me, came running over and wrapped his arms around me in a big hug. This impressed those present who were familiar with Japanese etiquette. They assured me that it was more usual for people to bow formally to one another, and that someone in a lowly position (as I was, relative to Kirk) was expected to bow lower than the boss. But I had at least a 30-centimetre head start on Kirk (quite literally), and I think he was pragmatic enough to have realised that country people don't go in for such formalities. He threw caution to the wind, and that was the beginning of a great friendship and mutual understanding.

Later in the trip, when there was a demonstration drive around the back roads from Raglan scheduled, Kirk and I were assigned to the same vehicle. He was pleased about that, and when we reached a very windy, shingle, ridge-top road that had been carefully selected to show off the ute's capabilities to a cast of motoring journalists and dealers, Kirk's eyes shone.

'Ahhhh!' he exclaimed 'Rally road!'

'Do you want to drive?' I asked him.

'No! More fun with you driving!'

Needing no further invitation, I leaned over and pulled his seatbelt extra tight.

'Do you get scared easily?' I asked him. 'Shall I give it a bit of a wind-up?'

The Japanese horror of *katanashi* — losing face — meant there was only ever one answer he could give. He nodded vigorously, and beamed from ear to ear like an excited child.

I had the green light. It hardly mattered what he replied, because I was just as susceptible to the sight of that wiggly, well-cambered, slippy-gravelled road as he was, and as soon as we were ready to go, the red mist descended. With just the right amount of weight in the back, the tyre pressure set the way I preferred, and having travelled that road every day for the past 10 days setting up the course, I felt confident enough to let rip. Off we shot.

The intercooled, turbocharged, multi-valve diesel donkey that the new model had aboard produced almost the same amount of power as an older-model 308 Holden V8, so it could move along smartly. I pushed the ute close to its limits. Using the technique of left-foot braking (instead of using your right, the accelerator foot) to enhance the handling and control, and flicking the transmission in and out of 4WD at high speed to control the drift on tight corners in the loose shingle, we managed to aggressively race past quite a few of the other motoring journalists' vehicles. Kirk, meanwhile, who was chattering excitedly at first, soon lapsed into silence.

With the rest of the crew left well in our dust, we soon faced a new challenge. We had pace notes, but they had been written in the winter, and the photos showing critical turn-off points had been taken when many of the trees were bare of leaves. Now the luxuriant spring greenery obscured many of the important marker points and we found ourselves a bit lost. As a back-up, if we failed to negotiate our navigational waypoints, we had a sealed envelope telling us where our mystery destination was, but if we opened it we lost the game. Unwilling to give it away that easily, we did much sheepish back-tracking and rechecking of photos, and Kirk somewhat recovered his composure.

When we eventually cruised into the luxury seaside accommodation — Mazda events are well-known for laying it on — we were on a great high. Kirk's latest baby, the BT50, had shown that not only was it a good off-road performer but also, in the hands of someone prepared to give it special treatment, it could serve as a pretty handy substitute for a rally car as well.

'We did not know the ute could do such things,' Kirk said in awe. 'We did not know it could go into four-wheel-drive at such speeds!'

Later, he sat me down at the table in his luxury room to show me photos on his laptop. Many had been taken on Waipari years before, and I was surprised at what they'd considered worthy of photographic attention — the old Holden Statesman power-steering system I had grafted onto our station Land Cruiser, using farm workshop odds and ends, for example. Some were plainly highly commercially sensitive, showing as they did some of the computer wizardry they used in design. I felt honoured to be shown them — but not half so honoured as I felt when he closed out of the photographs and there, on his desktop, amongst folders labelled *Mazda Japan*, *Mazda Thailand*, *Mazda China*, was a folder in the bottom corner labelled: *Mark, NZ*.

A few months later, Kirk sent me a bottle of sake that I am assured is well out of the ordinary, with a thank-you card that read: *A little bit of your ideas lives on in the new BT50.* Whenever I see a Mazda or Ford ute of that model (same thing, basically), I feel a deep sense of satisfaction, and I am just as grateful to Mazda for the opportunity to be a part of it all. The income I received from it helped me stay solvent through some difficult farming times, and, although I have had interesting jobs with some of the other motor companies, Mazda is number one in my heart.

Occasionally, I've been irritated over the years when someone has told me how lucky I have been to have managed to generate off-farm income at critical times. I've always been tempted to answer the way pioneering motor magnate Henry Ford answered when told how lucky he was: the harder I work, the luckier I become. But I have to admit that in the nature of some of the work I have had, where the jobs have ideally suited my loves and interests, I've been luckier than most.

We sometimes refer to the station as an overgrown sandpit, and love to welcome other like-minded people to share it with us, whether it's in off-road vehicles, on motorbikes, or horse-trekkers or hill-walkers. Our work tools and daily chores are often the same as other people's toys and entertainment! With the difficulties of vehicular access comes an exciting and spectacular 4WD challenge, the frustration of freeloading game becomes shooting-sport heaven, and an early-morning muster becomes an exhilarating horse or motorbike journey.

We have hosted many 4WD events at Waipari, from competition rallies, weekend 4WD trips and camp-outs to major 4WD fundraising treks. In 2004 I suggested I could run a one-day fundraiser for our local Waipawa Church. Much planning and debate went into ensuring a successful day, but one of my definite opinions, to ensure maximum support, was that the event should be held on Sunday, not Saturday, as was preferred by the church hierarchy. Much debate ensued around how that may affect church attendance. I was secretly very chuffed when the vicar announced that he agreed with me, and so would cancel all church services that Sunday! For years as a child I had daydreamed about going four-wheel-driving instead of being made to go to church! Such was the success of the event, and the attention of potential young parishioners at a 'church activity day', that the Bishop of Waiapu also took an interest in the concept. But my hopeful suggestion that the bishop make a decree stating that such ecclesiastical 4WD events should take Sunday precedence over normal church, while met with open-minded consideration, sadly never seemed to get formalised at Synod!

One of the other great events we began holding at Waipari was the Labrolympics down on the beach. As our beach neighbours Bruce and Jan Johnson, as well as Dave and Mary Clare Reynolds, all have Labradors, we decided to invite Marley, Merlot and Maisie's Labrador cousins to a day of doggy games on the beach. Events such as bum sniffing and biscuit-eating races, as well as 'dogs-leading-their-owners races', made for a day of canine hilarity! Medals were made out of manuka nuts (cut off round bits of wood we use to smoke meat on the barbeque with), while the flat-deck of the Land Cruiser was the winner's podium, and all the mess and doggy evidence was wiped out by the next high tide. But the stories and memories live on. There were also rumours of well-bred puppies as a bonus ...

CHAPTER 13

HIGH-VIS DISASTER

My accountant, Pita Alexander, occasionally tells me that if I have one foot in a bucket of boiling water and one in a bucket of ice, by the law of averages I should be feeling quite comfortable. It sounds fair, but it does ignore the problems that occur due to the extremes at each end. Farming is often like that, with the mild and benign days all too often bracketed by drought, or heavy snowfall, or floods. Being coastal, our bit of Hawke's Bay escapes the dangers of heavy snow that afflict the South Island high country where I served much of my farming apprenticeship. That's why, when I once said to James Innes on Haldon Station that I would love to farm in the Mackenzie Country one day — a favourite saying of mine is that my heart is in the Mackenzie Country, but my wallet is in Hawke's Bay — he immediately replied that he would swap Haldon for Waipari any day of the week. I was only a boy when he said this, and it made me think that maybe my great-grandfather had known what he was about when he selected land in coastal Hawke's Bay to farm. Similarly, Mark Acland, the wise and capable farmer who owned the magnificent Mount Somers Station, once told me that Hawke's Bay was a great place to farm.

But that's not to say farming our bit of paradise is easy. We have a traditional rainfall of around 48 inches (1200 millimetres) of rain per year. Most farmers would regard that as quite high, and if it were spread evenly throughout the year we would be (quite literally) in clover. But it all tends to fall in the winter, and not so much in the summer, meaning that we tend to get a lot of very deep mud in the winter, and often suffer drought conditions in the summer.

Drought creeps up on you slowly, and can be difficult to manage, as there is always the possibility and hope that good rain of a few inches will solve the problem overnight. The drought of 1983, my first year working on Waipari when it was under its previous management, was one of the worst we have known. The sights you see during a drought can be utterly heartbreaking. You often see lambs bogged in the deep mud of the rapidly evaporating dam where they are trying to get a drink — and sometimes they're drowned when a welcome rain falls and replenishes the dam while they are trapped there in the mud. You can see cows losing condition by the day as they try to find a mouthful of grass to turn into milk for their sucking calves. It's pitiful.

Then there are very heavy rains, caused by the water-laden easterlies that drag moisture in from the sea and then dump it as the clouds strike the rampart of our 500-metre coastal hills. When this happens, the sodden clay soils often can't hold the level of moisture, and gravity takes advantage, causing deep-seated slipping all over the hills. The silt debris washes down the engorged creek beds and eventually settles on the flats and in the estuaries. This process has been going on for as long as the land has been intensively farmed. Down the valley from us, there is a section of creek bank where you can see evidence of three separate fences erected, one on top of the other, over a period of about 100 years. That translates to about 3.5 metres of silt build-up in this valley in that time. In the Second World War, the Home Guard was instructed to guard the lagoon above Kairakau Beach — the estuary of the Mangakuri Creek — in case Japanese submarines tried to enter. Now I struggle to get my jetboat up there, especially at low tide, where there can be less than 300 millimetres of water over the mud. Some days you can walk across it without getting your knees wet.

In one sense, you could see this process as a good thing, as over time the steep hills and valleys will average out to become fertile flats. But that's the long game. In the meantime, we need to make a living out of harvesting fresh, high-quality grass on the existing hills and valleys.

Waipari was one of the early converts to the need to take soil conservation measures. Concerned about the silting of the Kairakau Lagoon, the authorities decided it had to take measures to control erosion in the headwaters of the Mangakuri Creek. Since most of these are on Waipari, it was decided to plant around 240 hectares of the station in forest, both to serve as erosion control and to provide an alternative land use. The planning was done with the enthusiastic support of Robin Black from the Hawke's Bay Catchment Board, and planting was one of the first jobs I was involved with when I came to work at Waipari under the manager in the winter of 1983. I didn't plant many of the trees: my main job was to carry the seedling trees and planting gang up in the Land Cruiser. As it was winter and the steep clay track was very slippery, the big, strong, tough and mostly gang-affiliated tree-planters were too scared to ride inside the Cruiser, preferring to hang off the outside ready to abandon ship at the first hint of a slide!

We continued to plant as much as possible during the downturn of the 1980s. While we received some financial help from the catchment board, we still had to find the balance and add it to the overdraft. But meanwhile, with forestry the beneficiary of a favourable tax break, offered as an incentive to plant more trees, it had become fashionable for Queen Street farmers to sink their money into forestry in order to avoid tax. This was never going to escape Roger Douglas's eagle eye, and he soon moved to close the loophole. This had the predictable effect of hitting those who were working hard for the sake of their land and the economy, at a time when everything else that Douglas and the Fourth Labour Government had done was brutalising us, too.

A few years later, I happened to be at an election-night dinner in the Hastings Club where Roger Douglas — Sir Roger, as he had lately become — was the guest speaker. After dinner, I was alone in the bar watching the election results coming over the TV. Sir Roger came and joined me. He had just formed the ACT Party at that point, and was openly pleased when a Labour MP was replaced with a National one, because he felt National would do more to support his financial reform crusade. It was fascinating to listen to him dissect the political strategies involved in an election campaign.

But I felt I had to express my displeasure at the forestry tax-deductibility changes.

Sir Roger looked me square in the eye. 'You farmers weren't the ones we needed to target. Don't tell me you didn't find a way around it?'

In fact, as it turned out, we were planting trees for the initial purpose of soil conservation, rather than just tax-avoidance investment. He was right.

I admired his openness, but it was a subject on which we had to agree to differ. We had done it tough. I had set up an investment proposal whereby others could invest in our forestry company, with an indicative rate of return on investment of about 12 per cent. This attracted some interest, but my timing was poor. It was just after the 1987 sharemarket crash, and persuading investors to part with their cash was like trying to catch cockabullies with a rake. In the end, it was the death of my grandfather in 1988 that bailed us out of a financial hole. I received a small legacy. I could have put that money towards a nicer car, as so many would have done. Instead, I put up with a $200 Valiant that had so many holes rusted in the boot that I didn't put tools in there for fear of losing them on the road. Instead, I used the inheritance to cover some of the immediate forestry costs, lending the cash to the partnership to enable more planting.

It has proved to be a stellar investment. It cost about 15 cents (nowadays 50 cents) to put a pine tree in the ground, and 25 years later, for a few extra dollars spent on tending and pruning, that tree was worth about $150. If they were in the pasture, I calculate that the forestry blocks would carry less than half the average livestock rate (about 10 stock units per hectare) of the rest of the property. To put it in perspective, one of the first blocks to be put in forest — the Upper Nikau Block, 70 hectares of land so poor that the manager at the time pronounced it too infertile even to grow pines — lately returned almost double what the government valuation on the whole 1300-hectare station was when the trees were planted!

And we have reduced considerably the soil erosion and improved the quality of the water running down Mangakuri Creek to Kairakau Beach.

All the same, I've never bought myself a brand new car. I have an aversion bordering on a phobia with regard to rapidly depreciating assets.

And it's hard to put a price on what the forestry blocks deliver to us in their other role — namely, soil conservation — when it really rains.

It so happened we were just in the process of harvesting some of our trees in 2011 when a bunch of different weather systems joined forces out in the Pacific to create what meteorology types refer to as a 'weather bomb'. What that meant for us was rain, and rain in almost biblical quantities.

On the first day of the deluge, I was hiding away in my hilltop man cave (more of which later) doing office work. The rain beat incessantly on the roof, but I didn't really give it much thought until about midday, when I had to attend a meeting in Waipukurau. As I was going down the hill road from my elevated perch, I was astounded by the sight of slips on the adjacent hillsides, slips everywhere, visible through the misty curtain of rain as vivid yellow slashes in the grey-green grass. When I reached the road proper, it was more like a river, with dirty water coursing along it, carrying mud and vegetation. I eased into the filthy water and drove up-river slowly and carefully until I suddenly felt the ute's front axle drop. I knew that it was a distinct possibility that the carriageway could be washed out, and if I hit such a bite out of the road, the water would easily float the ute and carry it away, with me in it, and no one would be any the wiser until my body was found a few days later. I carefully reversed back out and retreated back up the hill to come down cross-country, an exercise more in vehicular skiing than in driving. As long as you keep the front wheels, which do the steering, turning a bit faster than the ground rushing past you, you can maintain some limited form of directional control.

The bottom gate was choked with silt and trees. When I stepped out of the ute, there was fully 300 millimetres of swirling water over the spongy ground, but I was able to lift the gate off its hinges and drive through onto higher ground 100 metres away. From here, I could just make out the top of a 6-foot culvert in the stream. That gave me the confidence to drive over it, and soon I made it to the homestead.

My son Henry was there. I told him we were heading into town, as I reckoned it would be great experience for him. He would learn far more about driving through flood waters if he came along than if he heard about it from me relating it like some kind of legend later on.

The local shingle road was inundated, but I deemed it passable. The local rule of thumb was that if you could still see the tops of the fenceposts and were in a proper 4WD vehicle, it was safe to drive through flood waters. It helped if it was daylight, because then you could see beyond the deeper bits to where the roadside markers protruded from the water. This gives you an all-important visual reference. Otherwise, it's too easy to be disorientated by the water rushing past sideways, so that you drift diagonally relative to your desired course and drive off the road. With the ute in low gear, we drove slowly along, my eyes fixed firmly on our exit point, Henry staring apprehensively out at the water that was lapping nearly level with the bonnet.

Once we reached the tarseal road, things became a bit easier, and as we headed further inland we climbed out of the worst of the flood altogether. At the top of one hill, we saw a police car heading in the opposite direction. I rolled down my window as we came abreast, and he did the same.

'What's it like up your way?' he asked. 'Will I get through?'

I shook my head.

'Not in that,' I said, nodding at his Holden Commodore. 'You'll probably get as far as the Omakere Hall, but you'll struggle after that.'

He nodded.

'Appreciate that. Might be a good idea to head back to Waipawa.'

We left him to do a U-turn and follow us.

In Waipukurau, we dropped into the local newspaper office. Henry had taken some photos on the drive in, and he downloaded them to their computer to run in the following day's paper. I went to my meeting and then, after we'd done a quick grocery shop, we turned the ute for home. It had been raining all this time.

Before we even reached the point where the tarseal gave out and the shingle started, we saw a Road Closed sign, standing forlornly in the darkening shelter of a stand of macrocarpa trees. We politely ignored it. I stopped, engaged low-range 4WD, and we prepared to do battle with the muddy water that ran as deep as 500 millimetres over the tarsealed surface. We crawled along a road at walking speed where we would normally do 100 kilometres per hour.

Still the rain fell, and darkness was descending, too. Turning onto Clareinch Road, the last stretch before home, we needed the high-powered off-road driving lights mounted on top of the bull-bars to see the way ahead. The stark white illumination showed lazily eddying and swirling water that was plainly much deeper than it had been a few hours before.

Using a powerful handheld spotlight that we normally used for night-shooting rabbits and possums, we managed to pick out a road marker about 200 metres distant. I breathed a sigh of relief. With a visual reference, we could risk the crossing. With Henry keeping the beam steady on the marker, we eased our way slowly but surely into the black, swirling waters.

'There's a lamb floating past!' Henry exclaimed.

'Keep that light on the marker,' I told him. 'No time for counting dead stock now.'

I expected this stretch of flood water to be the worst, and as we emerged from it I half-expected the ute to set to and shake itself as a soggy Labrador does after a swim in a dam. Sure enough, apart from big shingle banks deposited in the middle of the road by the rushing water, the rest of the trip was more about dodging downed trees than deep water. Realising that I'd never make it up the slippery track to the man cave in the dark, I stayed the night in the homestead at the bottom of the hill, listening to the rain bombarding the roof and thrumming in the gutters.

As soon as it was light enough to get anything done, I ventured out again. I changed my choice of wheels, opting to take the 4WD tractor instead of the ute. The water seemed to have subsided somewhat, and I used the front-end loader bucket to push some of the logs off the road and to divert some of the worst of the torrents in their newly-formed stream beds away from the roads.

I was only just beginning to get a sense of the scale of the flood and the damage it had caused.

Huge trees had been uprooted and had been piled up against our bridge, and the flood waters had gone clean over the deck to a depth of about 3 metres. Every fence within 20 metres of the creek had vanished, and about half a metre of silt seemed to cover absolutely everything.

The Power Board were anxious to start the mammoth task of restoring and rebuilding the network, and the police were keen to evacuate families stranded at Mangakuri Beach, but everyone was stuck until the road and bridge could be cleared. Luckily, the crew who were harvesting our forestry block still had their machinery parked on the correct side of the creek. When the boys showed up to work, these proved immensely valuable in helping to reopen the road. One 30-ton log-picker crawled its way up the road, and plucked fallen trees off power lines and shifted stray logs off the roadway. Another excavator clawed at the log jam threatening the bridge, while the log-skidder went on down the road pushing off logs, in an effort to push a path through the worst of the silt. I followed the skidder with the tractor and the grader blade, but it was a bit like Moses trying to part the Red Sea. As soon as I bladed the gloppy silt to the side of the road, it slumped back and reclaimed the track behind me — much to the frustration of the Centralines team, who were forever hooking a tow rope to their ute so I could haul them out of the slippery brown mess.

By that afternoon, I was able to get my motorbike across the bridge and attempt to survey the damage. There were plenty of helicopters and planes flying about, but they were intent on sightseeing (and in some cases made things worse by chasing stock further into bogs). I was left to survey the damage from the ground.

I could only go about a kilometre up our previously well-formed forestry road before I was confronted by huge slips right across the carriageway. We still had about 20 truckloads of logs stuck on a forestry skid beyond the blockage, and they would lose value dramatically if we couldn't get them out within a week. Numbed by the scale of the damage, I retreated and found a way up the other side of the valley to where I could get cellphone coverage. I called a good mate, Mike Walmsley, who owned a large contracting company in Hastings.

'Mike, we're in a hell of a mess down here,' I said. 'I don't know if we need dozers or diggers.'

'I'd better take a look,' he replied without hesitation. 'I'll be down in an hour.'

True to his word, he arrived an hour later with his Honda XR600 enduro bike on the trailer behind his vehicle, ready to reconnoitre and plan the rebuild. Mike is an accomplished rider, and I did my best to stick with him. To say it was challenging and eventful would be an understatement. Against your instincts, you have to be prepared to use lots of power to keep your bike upright in the slippery, treacherous conditions that prevailed. We both had a few spills, but one of us a few more than the other ...

We were coated from head to foot in sticky mud after a few hours' reconnaissance. We retired to the man cave for a wash and to plan our next move over a very welcome beer and a lump of steak.

'A digger will be best for this job,' Mike pronounced, and he jacked up his best operator, Ian Ericson, to be there with a 20-ton Caterpillar digger at first light. We celebrated this decision with a bottle of Appleton's rum, while browsing the flood damage pictures that were by now thronging the internet.

Next morning, at first light as promised, Ian arrived with the digger on its transporter.

'I'm sure they meant well,' he was saying, 'but they had a roadblock just up the road and they reckoned anyone going past had to have a bit of paper from Civil Defence HQ back in town. They asked if I was going to do a council job, and when I told them the digger was for Waipari, they told me to turn around and drive back to town to pick up the official piece of paper.'

'So what did you say?'

'"Like hell," I told them. "Either you move your roadblock, or I'll unload the digger and do it myself!!"'

So here he was, with the digger and without the bit of paper. Ian's not a man to trifle with. A little later on that day, I heard that members of the Pan Pac forestry crew who were doing our harvesting were being turned back at the roadblock, which was being manned 24 hours. I made a very fiery phone call to the Civil Defence controller, which saw the Road Closed sign moved to a point just past our station yard. For the first few days, we drove off the road and around it. Then we decided that was pointless, and moved the sign to the side of the road. The poor bloke charged with keeping the sign in the right place was told to keep well out of my way. It's the only time I'll admit that an OSH issue may have existed — his safety may have been in jeopardy!

Civil Defence later told me that since the phones were out of order and they couldn't get an on-the-ground update from the locals, they had unilaterally decided the roads were impassable.

I pointed out to them that I was on the rural volunteer fire brigade and they had my cellphone number. Why hadn't they used that?

They didn't have a worthwhile reply to that one.

While they stuck pins in wall maps and procrastinated, no doubt filling in hazard-awareness plans and upgrading the OSH manual on how to keep safe in a flood, it was our little team of locals with our own gear that got the road open. It was the turbocharged, distilled essence of bureaucratic, power-crazed, clipboard-toting, hi-vis-wearing, hard-hatted incompetence.

The stress of working 16-hour days trying to restore essential access, perform necessary farm jobs and make repairs such as reinstating house water systems, all while dealing with constantly stuck vehicles and overly friendly clay that just won't let you go … all of it did start to take its toll.

One night at around 9pm, when I'd already been working for 14 hours, I was driving the 20-ton digger to another job down the road. I had all the lights on but, as they were mostly blotted out by sticky yellow clay, I was finding my way more by instinct and the moonlight. Suddenly there was a brilliant flash, followed at once by another.

Strange to have lightning, I thought, on such a fine night.

Then I realised the digger boom had collected the live power lines above.

Instantly the adrenaline kicked in. I stopped and carefully reversed. To my relief, I managed to extricate the machine from the wires. Realising how tired I was, I parked the machine in as safe a place as possible, jumped out and walked home for a good night's sleep.

'You're off-grid up there, aren't you?' one of our team asked me the following day.

I nodded.

'Lucky bugger. The power was off in the whole valley last night.'

'Oh?' I said.

'Yeah. The Power Board came and had a look, but they couldn't find the reason. Weird, eh?'

I nodded again. I hadn't brought any wires down: luckily, the fuses must have burnt out just in time, and there was little evidence of my mishap besides scorch marks on the boom of the digger, safely hidden out of sight. I muttered a quiet apology to Centralines; I possibly owe them a box of beer.

The gold medal for the wasteful application of bureaucratic red tape came a few years later when we were proposing to perform far more routine forestry operations. The council insisted on a week's notice and a hefty fee so that they could form a traffic-management plan and assign a couple of officially trained and suitably qualified and dressed lollipop operators so that the same log-picker could pick up a few logs off the roadside in a very similar place. Anyone driving up to the stoppage would be stuck there for 10 minutes before they got the all-clear to go, and it would cost us a few thousand dollars. Needless to say, I wasn't allowed anywhere near the operation. The officials were doubtless terrified that I would revert to the Charlie Upham method of just getting the job done, and bugger the brass. And nor did they get to charge me, as they initially indicated they would. Someone else paid that bill, in a wise attempt to keep the peace.

As we were working to make good the flood damage, the then stock manager, Barry Andrews, pointed out to me that the fences and flood-gates at the foot of every paddock were gone — except on the forestry blocks, where the trees had evidently slowed the destructive flow of the water to a level where the fencing infrastructure could cope. That was proof of the worth of the trees in erosion control alone. Better still, the proceeds from the 60-hectare block we successfully harvested in and around the flood more or less covered the cost of the repairs to tracks, pasture and fences — $250,000-odd — that the flood obliged us to do. For others in the district faced with similar mitigation bills and no option but to borrow to meet them, it was a devastating blow.

So often, farmers tend to think that growing grass to feed livestock is the only way to be a successful farmer. But forestry has proved to be an excellent supplementary business to our pastoral farming activities. In fact, it has probably enabled us to carry on farming sustainably. Over the past 30-odd years, I have arrived at the opinion that any block of land bigger than about 10 hectares that won't carry more than five stock units per hectare should be fenced off and planted with protection/production forest. Where they're steep and rough, these blocks aren't worth fertilising, tend to be the hardest places to muster, and end up being the places where stock go to get lost or die anyway. If you take these areas out of the intensive farming systems, you can concentrate on improving and developing the better country — and at a pinch you can still graze them in times of need, and treat them as rough feed banks.

We have turned about 25 per cent of Waipari over to forestry, but, with the improvements in fertiliser, fencing and irrigation we've seen along the way, the carrying capacity of the remaining 1000 hectares hasn't dropped very much at all.

An ambitious young man — he reckoned he was just a poor student — recently asked me if I knew an easy way to make a million dollars. (Lucky for him, he didn't add the word 'quick'!) Doing a swift mental calculation, I told him to go and find some suitable land and plant and nurture about 10,000 good, well-pruned pine trees. About 25 years ago, each seedling pine tree cost about 15 cents. So for about $1500, plus the cost of the land and a little more for the labour required to plant, release and prune them, you could get a very good return on the investment of your low-interest student loan.

Money may not grow on trees, but it grows in them!

CHAPTER 14

REALISING THE DREAM

If there's one thing that farming teaches you, it's that there is no constant but change. You can make changes all you like, but even when you're happy with everything — perhaps even especially when you're happy with everything — change can overtake you. Like four-wheel-driving, the trick is to maintain at least the illusion of control over the way things are unfolding.

As the foregoing chapters record, the time since I took over Waipari has seen some significant changes — in New Zealand, in my corner of it and on my farm. But there have been others that I haven't set down.

Before my Grandfather Alwyn died in 1988, I never got to sit down with him and hear whether he was proud of, perhaps even impressed by, what I had done to secure the family legacy. Perhaps, being the kind of man he was, he wouldn't have let on, anyway.

In mid-2005, I was at the Patangata Tavern, south of Havelock North in Hawke's Bay, enjoying a district get-together. Amongst the group was National Party stalwart, long-serving MP and sometime Cabinet Minister John Falloon and his wife, Phillipa, who were on their way to stay with me at Waipari after a long day on the campaign trail. John was on the piano. Those who knew him will remember he was an incredible jazz pianist, who could (and would) jump on any available piano and play by ear anything anyone asked him to play.

In the middle of all that, I got a phone call.

'Mark?'

It was Mother. I had to plug my ear and lean into a quiet corner to try to hear what she was saying.

'Sorry. What was that again?'

'Your father has passed away.'

After I had finished talking to her, I went outside to be alone, as you do when you've just learned of your father's death. I stood under a tree with the sounds of the merry crowd and the piano drifting from inside and tried to make sense of it all.

When I'd got it all more or less in perspective, I went back in and re-joined the party. It was only when it was winding up and we'd said our goodbyes to the rest of the team that I asked John if he could play 'Stairway to Heaven'.

'Of course,' he said, without a moment's hesitation. 'But why?'

'My dad just died an hour ago,' I said. 'I reckon it'd be fitting.'

He played a beautiful rendition of the song, and then we headed back to Waipari. John and Phillipa sat up with me until the wee hours, and we drank a few toasts and they let me ramble —

as you do, at such times — about all kinds of stuff. We talked a lot about how we had survived Rogernomics. John had been Minister of Agriculture and Forestry in the National Government that had succeeded Douglas's Fourth Labour Government, so he had a fair bit of perspective. When the mood was getting a bit sombre, we cheered ourselves up by comparing bulldozer frights and tales of the odd things we had done to make ends meet.

'You've got to write a book about all this stuff,' John told me, wiping a tear from his eye.

Six months later, John himself was dead: a brain tumour for which he had previously received surgery having returned. I had the honour of being amongst those who shovelled soil into his grave. I would have fired up his Caterpillar bulldozer, parked nearby, to do the job, but perhaps decorum was more important than efficiency in the end.

So one generation passes. But another generation was rising. My marriage to Chris didn't last, but it produced my beautiful daughter, Emma, in 1988. Chris and I later separated. She was a hard worker, and really put the hours into the farm, as we all had to back then if we meant to survive. But it took me some time to realise that we shouldn't be married.

So eventually we went our separate ways.

A while later, I married Sue.

Sadly that marriage didn't last, either. Still, one thing for which I owe a debt of gratitude to Sue is that her schoolteacher training led her to suspect I was dyslexic. So at the age of 40, I submitted myself to a set of tests, and her suspicions were confirmed. A whole lot of things suddenly made a lot more sense to me — the way words had always confounded me, shifting around on the page and in my head like a mob of unruly ewes, and the way I occasionally seemed to see the world a bit differently to everyone else.

In our 10-odd years, we had three wonderful boys together — William in 1996, Henry in 1997 and Jack in 1999. It's been both amazing and slightly baffling watching them all grow and flourish, and seeing their different personalities emerging. Of course, being a firm believer in learning life skills on the job, there have been moments. Some come easily to mind. One day, as an after-school treat, I had five-year-old Jack in the cab of the tractor with me while I was clearing out the water table around a culvert on the side of a farm track. When I bought that tractor — an 80hp Deutz-Fahr, which, with its very low centre of gravity, four wheel brakes and twin diff locks, was the best steep hill-country tractor then on the market, in my humble opinion — I had stipulated that there should be an extra seat (with seatbelt) in the cab where an instructor could sit. That way I could have a small boy with me, safe in the knowledge that so long as he was strapped in, he was safe in the cab.

Good job, too. Even a tractor that sits as nicely on the ground as the Deutz does can become unstable when you hoist aloft a front-end loader bucket full of a half-ton of wet mud. I can't have been paying quite enough attention — there's nothing like having an excited five-year-old peppering you with questions

about which lever does what to take your mind off the job —
because when I reversed to dump the load, one wheel dropped
into a hole and the other rose over a bump and the tractor began
to tip over. I wasn't unduly alarmed. It was more in the nature
of a gentle flop than a roll in anger, but Jack was on the top side
where the whole manoeuvre was accentuated. He screamed, a
look of absolute terror on his face.

In the split-second available, I tried both to drop the bucket
to lower the roll point and to hold Jack in his seat with the
elbow of the same arm. But he was wriggling and scrambling to
save himself, and my arm was squashed against the main seat.
I couldn't get the bucket down in time, so over she went.

Despite the tractor lying almost on its side, the oil pressure
stayed normal, so I was able to keep the engine running and, by
locking both diffs and using the down pressure of the loader,
I was able to edge the tractor backwards and push it back onto
its wheels.

'What was all the fuss about, mate?' I asked Jack.

He stared at me for a moment, then proceeded to tell me off
for giving him a fright.

It was all part of our controlled-crash programme, by which
our boys have learned to identify where the limit is and to be
better equipped to operate within it. No wrapping them up in
cotton wool in our operation!

We applied more or less the same approach to teaching them
to drive, and, as is the way of things, it didn't always go smoothly.

One late Saturday afternoon in winter, I was on heifer calving
duty, which meant checking the rising two-year-old heifers that

were due to calve and shifting their break fence to give them another pre-birth café meal. William, who was 12, walked with me about a kilometre down to the calving paddock, and on the way we enjoyed some quality father-and-son time, during which I explained to Wills why we were doing the job and how it was best done. He seemed in a particularly receptive mood, so I decided to take the tractor, which was parked in the paddock, back to the shed and to give him the chance to drive. He was keen! So I parked myself in the instructor's seat beside him and we set off.

He managed the first two gateways and got across the bridge without any alarms or surprises, but as we came into the tractor yard and lined up the entrance to the shed, I was a bit concerned to see Wills trying to stop the tractor by pushing on the brake without also putting his foot on the clutch.

I could see an 'unplanned incident' unfolding. I reached around Wills to pull the red fuel cut-off knob to stop the engine, but as the front-end loader forks made contact with the back wall of the shed, Wills realised things were getting out of hand, too. He jerked backwards, knocking my arm away. The engine had been spluttering to a halt, but now roared back into life in the split-second it took for me to make another grab for the cut-off. This was just enough to force the loader through the back wall, and there was a crash as the shelving on the other side of it, laden with carefully stacked fencing materials, toppled over.

The engine died, and Wills and I sat for a moment in the silence. I saw his face, contorted with terror.

'Wills,' I said gently, 'you're meant to put your foot on the clutch.'

'I thought you were going to do that part, Dad,' he replied.

Some problems are solved by applying more light, and some by applying heat. I decided this moment called for more of the former. I forced a smile and said: 'No, Wills. You were driving. It was your job!'

As we climbed out of the tractor, he surveyed the damage he had caused, clearly mortified.

'Don't worry, Wills. No one got hurt.'

I put my arm around him and we set off for home in silence. After a few minutes, I said: 'Wills, it's only a problem if we don't learn from the mistake.'

I let that sink in for another few minutes.

'So what did we learn from that?' I asked.

He was quiet at first while he struggled to frame the words.

'That Henry should do all the tractor-driving!' he blurted.

We agreed not to tell his mother for now.

But the secret soon leaked out, and the next morning, Henry (who was about 10) was raring to go when I asked who was going to help me check the heifers. He was probably more interested in seeing the legendary wrecked shed than the job itself, but I let him tag along.

Just short of the shed, we both stopped. I stared in disbelief. The rear wall of the shed was in place and there was hardly any evidence of the mishap at all.

'See?' I said to Henry. 'It wasn't as bad as Wills made out.'

Henry made no attempt to disguise his disappointment.

It turned out that our employees Dessy and Carly and their family had heard the accident and had mobilised to fix the shed early that Sunday morning. Dessy and Carly had both been unemployed when I had offered to interview Dessy for a job earlier that year. An indication of how keen he was is that Dessy was waiting at the gate at 7.30 so as not to be late for the 8am appointment. Although he was only in his mid-thirties, Dessy had nearly enough children to justify our own school bus. To a lot of people they may have looked like unpromising material when I employed them, but Dessy and Carly were real treasures. In their time with us, they repaid the leap of faith we took with them many times over. Quite apart from their excellent work in the routine business of the farm, Carly was a big help when it came to renovating the homestead. It was a pleasure to be able to give them glowing references when the time came for them to move on to another local job, and a joy to see them succeed just as well in that one, too.

※※※※※※※※※※※※※

'Dad! Dad!'

There was a note of urgency in Jack's voice as he burst through the homestead door.

'Is the new tractor well insured?'

That definitely got my attention.

'Why do you need to know?' I asked. 'What's happened?'

Worry verging on panic creased my 17-year-old youngest son's face.

'I'm afraid it's in a bit of a tricky position,' he grimaced. 'It might roll over the bank down into the creek. I might have chosen the wrong crossing, and when I realised it was a bit dangerous I couldn't back up because I had the harrows on. It is insured, right?'

I admired the fact that he was giving it to me straight. At the same time, I was thinking about the mostly flat paddock he had been working and trying to think of a spot where he could have got into any serious kind of trouble. Nothing sprang to mind. It couldn't be that bad, surely?

We hopped into the Nissan Patrol and headed directly to the paddock in question. As we got close, I saw that he hadn't been exaggerating in the slightest.

The tractor had a set of tandem discs and a double set of heavy harrows behind it, and it was halfway down a track that cut 30 metres from the flats to the creek bed. A slip had taken a big bite out of the track, and what we were contemplating was a 3-metre-wide tractor on a 2.5-metre-wide track. One set of dual wheels were hanging out in space, and the machine was teetering at a 30-degree angle.

I realised that I hadn't allowed for my son's tendency to take advice somewhat literally when I had told him that the tractor was very stable and you'd practically need to drive it off a cliff to get it to roll. He'd done everything but! All the same, I was impressed with how he had responded. He had had the presence of mind to turn the tractor off, pull the brake on as hard as possible and run home when he had realised he was in serious trouble.

Studying the situation from every angle, I decided that if the harrows were removed, I might just be able to ease the tractor carefully back out. We hitched the Patrol onto the harrows and hauled them back out of the way. Then we attached the Patrol to the back of the discs, the idea being that Jack would take the weight of the discs using the Patrol, and I would try to ease the tractor backwards from its perch on the slip. I made sure Jack understood every step of the operation, especially the instruction to make sure he had his seatbelt on tight. If the tractor looked like it was slipping over the edge, he wasn't to jump out. He was to hang on tight and dig in as hard as possible to help anchor me from rolling off the slip and dropping 7 metres into the creek below.

I gingerly approached the 5-ton tractor and rocked it experimentally by hand to see if it would move.

It didn't.

I climbed in carefully, and quickly fastened the seatbelt as tight as possible.

Starting the engine and triple-checking it was in reverse, in the right ratio and in 4WD with both diffs locked, I gave the call over the RT for Jack to take the weight while I gently tried to reverse.

As soon we applied power, the whole tractor started to slide sideways over the edge as the bank below stated to give way.

'Full pull!' I called urgently to Jack, but I knew it was going more sideways than backwards.

'Hold it tight!' I yelled, even as I braced myself for the inevitable slide and roll upside-down into the creek below.

But after lurching about 30 centimetres sideways, the tractor steadied. The bank seemed to hold.

I squeezed the TX button on the radio telephone.

'Mayday! Mayday! Mayday!' I said. I had never made this call in my life before. But then again, I was pretty sure I'd never before faced a situation that required all available help as soon as possible.

The RT crackled back immediately.

'What's the situation, Mark?'

It was Nigel Hanan, the Waipari manager.

'I'm in a tight spot in the tractor,' I told him. 'Can you get Ed to come as quickly as possible with his big John Deere and the strongest ropes he can find?'

'Roger. Will do.'

I sat with my feet hard on the brakes, watching out for the sensation of the ground crumbling beneath the tractor and ready to brace for the impact. The adrenaline was pumping to the max. I was as worried as I'd ever been. I was prepared to ride it out, even if it meant dropping headfirst into the creek. As much as every instinct screams at you to get the hell out of a vehicle in a situation like this, I was well aware that most fatalities in roll-over accidents happen when the person tries to get out, or is thrown out due to lack of restraint, whereupon they get squashed as the vehicle rolls over them. As I have said before, your best chance of survival and of escaping serious injury is if you stay tightly belted inside the cab.

But after I had sat there for a few minutes with no apparent change for the worse, I asked Jack to try to rock the tractor with all his might to see how stable it was.

It didn't move.

I thought that perhaps I could undo my belt, weave my stiff knee — I had recently undergone the knee replacement — through the various control levers and jump quickly before it rolled over the bank.

Double-checking that the exit was as clear as possible, I asked Jack to be ready to pull me way from the tractor by any means possible if it started to roll over or slide again. Then, as calmly and deliberately as I could, I sprang my belt, wriggled clear of the controls and jumped as hard and far as I could. I landed on firm ground — much to the relief of Jack, who thought he was going to see his father seriously injured before his eyes.

My heart rate started to slow down.

Soon we heard the roar of a diesel and saw the welcome sight of our neighbour Ed Bell on his 120hp John Deere tractor hurrying across the paddock. We soon had it hitched up, Ed's green tractor to my red one, as an anchor to prevent catastrophe, but then we were faced with the much trickier problem of how to recover it.

Logically, the tractor needed solid support to stop further sideways slippage. I figured the best way to get that was by attaching the 8000-kilogram recovery strap securely to the back axle and getting the John Deere to pull at a 45-degree angle to stop the tractor rolling over further.

I reckoned the odds were about 50:50 that we would save the tractor — or lose it over the bank into the creek. We needed someone in it to do the driving, and I asked at the planning meeting we convened before we started the operation who wanted to do what job. Ed preferred to drive his own tractor.

Nigel was pretty clear that he wasn't going anywhere near our tractor, and that he thought I (or anyone else) would be mad to get back in it. But someone had to. Ed climbed onto his big tractor. Jack had the Patrol to act as an anchor, and Nigel sat well back to co-ordinate the effort with his RT. I put on a good helmet (it's only your head that has the freedom to take a beating when you're firmly strapped in and a vehicle goes upside down), and gingerly climbed back into the cab. I felt a bit more comfortable with Ed and his tractor acting as an anchor. I gave the instruction to keep pulling, no matter what, once I got my tractor moving, and told Ed that he should keep driving straight through the fence if he had to.

Nigel orchestrated a successful extraction manoeuvre, Ed's 5-ton John Deere dropping its shoulder and pulling with all its considerable might, and the Patrol belching black smoke as Jack dug into every reserve of its power so that it took the maximum possible share of the weight. My slightly dodgy heart outdid all the machinery, racing at about 140 beats per minute as I eased the clutch, carefully ensuring we were moving more backwards than sideways from the rapidly crumbling bank.

Within three minutes, it was all over. Ropes were coiled up and Jack was firmly instructed to get back to work — perhaps avoiding that crossing in future.

***

Hiccups such as these aside, we've managed to raise the kids into adulthood.

Two failed marriages had made me understandably wary of serious relationships, but eventually I allowed myself to be won over again. After a five-year romance, Julie Holden has become a vital part of my life, well surpassing what former All Black and media personality Marc Ellis once described in his book, after a visit to my man cave, as 'an honorary bloke allowed in after dark or, if really lucky, at lunchtime'. Julie came with her own, ready-made family, which we have successfully blended with my own. Everyone makes themselves useful, in one way or another. Between us, we have two Emmas. My Emma is a caring, sharing one, always looking out for everyone. Emma Holden is a foodie, which means that family gatherings are very well catered. Her brother, James, worked until recently in the Canterbury Crusaders' strength and conditioning department (he's with Northland Rugby now), and he was helpful to the point of being cruel when I needed rehabilitation after the rebuild of my knee. Julie's youngest, Lucy, is keen on farming. She did very well in a stint as shepherd-general on Waipari over a recent summer gone, and at the time of writing she is acting as big sister to Henry at Lincoln, where both are doing degrees in Agricultural Commerce.

I have enjoyed watching my own children's talents emerging. William is our computer helpdesk. He's got a real aptitude for the infernal things, setting up and maintaining them. Jack is a master cameraman, and the rest of the family defers to him when special events need recording. Most recently, he's dragged us (or some of us, anyway) screaming into the twenty-first century with his expertise in aerial drone photography.

You see their different personalities in the way they approach and handle vehicles. Although they all learned in the same steep and slippery conditions on Waipari — excellent for teaching vehicle control, if not necessarily the finer points of the Road Code — they all come at it differently. Emma is careful and gentle. She chooses the safest set of wheels on the place — the John Deere Gator is her favourite — and she carefully plans her trip and makes sure everyone has their seatbelt on.

William tends to treat vehicles as a means to an end, the end usually being his next academic assignment. He carefully checks everything and analyses the various mechanical and electronic features of any vehicle he climbs into, in order to satisfy himself that he's utilising the full complement of gadgetry.

Henry, on the other hand, will likely have dismantled and modified the vehicle to make it go a bit better than before. He's the one most likely to exceed the vehicle's limits, but also the one I'd back to salvage the position singlehandedly when he does.

Jack will jump in, fully expecting his brothers to have done all the required maintenance, and roar off to set speed records. If there's a motorbike race involved, Jack will move straight to the front, no mechanical mercy shown. But he's quite likely, in the act of looking around to make sure he's still in front, to fail to see a hole in the ground ahead! For all his impetuosity though, Jack has a delicate and measured touch on the controls of a digger or dozer — a reflection, I suppose, of a youth misspent with a game controller.

My living arrangements have changed, too. In 2008, when Sue and I came to a parting of our ways, we faced the problem of who should live where. The conventional method of solving this situation would have seen Sue shifted out to a house in town, but that wasn't satisfactory. She wanted the boys to live with her, and I didn't want them to lose the country lifestyle I had tried so hard to provide for them. So it looked like I would have to be the one to move.

There was a particular vantage point in our coastal block which gave a view over the station in one direction and the beach and the unbroken horizon in the other. I had always harboured dreams of building and living in a basic hut perched here — something like a modern version of the Craigs' hut in the movie of *The Man from Snowy River*. With the perfect excuse, and with the help of good friends Johnny and Rose Roil, whose Hastings business specialises in country cottages, I set about designing it.

The brief was simple. It had to be centred around a decent wood fire, fully off the grid and self-sufficient, offer the best possible view of the station and the beach, and be accessible only by 4WD track, to maintain privacy. It was pure good fortune that my chosen site was bathed in a little ray of high-quality cellphone coverage, because this enabled me to have a high-speed internet connection. That meant I could run my businesses from there as well.

Designed to withstand wind loading of up to 200 kilometres per hour, the building has two bedrooms, an indoor loo and shower, as well as a bath and a shower on the back deck.

Water is collected from the roof and heated by solar collectors, photovoltaic panels provide electricity, and a Kent Barker log fire with a super wetback does for cooking and space heating. It's fully self-sufficient. In honour of that fact, and the fact that it was largely paid for with carbon credits from our forestry activities that I cashed in at close to the top of the market, and with all the talk about carbon footprints and so on, I was going to call it No-Carbon Cottage. But someone christened it the 'man cave', and that name has stuck. In many ways, I am living the Kiwi bloke's dream — shooting clay birds from the deck, wallowing in the solar-heated bath water while watching cruise ships glide by, choosing from a range of different barbeques to use to cook our very own, world-famous vegetarian lamb chops when the weather is conducive. ('Vegetarian lamb' refers to lamb fed on naturally growing grass and herbs, rather than grain.) It was pretty spartan at first, but Julie's feminine touch has worn the hard masculine edges off somewhat. Where once the shelves were cluttered with tools, ammo, motoring magazines and plenty of dust, they are now orderly and there are flowers, assorted bottles of nail polish and *NZ Life & Leisure* magazines rubbing shoulders with the petrol-head titles. I have to admit, it's better this way.

As soon as it became generally known that I had built the man cave, people started wondering how I had got planning permission. I'm usually a bit coy about answering this, because I followed a very good rule that John Lee of the Cardrona Valley once taught me: it's far easier to beg forgiveness than ask permission. Same deal when I discovered I couldn't quite see the waves breaking on the beach from my pillow because there was a small hill in the way. Three days' work with a D6 bulldozer fixed that problem, without any need to trouble the bureaucrats. All this, and I got the house for less than the price of a Range Rover!

When I bought a parcel of neighbouring land that was part of a farm park development above the beach, it came with access rights down to the private beaches below the man cave as well as the use of the all-weather road. But it was much simpler and quicker to improve the 2-kilometre track up and across the paddocks, and while you have to have a 4WD to negotiate it, that serves to keep out a lot of the opportunist door-knockers we get in the country trying to sell us things or save our souls. Now they just leave me a letter in the mailbox 3 kilometres away at the woolshed, saying they tried, but failed, to find my house. I'm pretty relaxed about that. I don't have an address; I have GPS co-ordinates instead.

Most of the time, Julie and I live in the man cave, loving being at one with Nature, watching the sea's many moods out one side and the sun setting over the Waipari hills out the other. We don't have to worry about what we wear to work each day unless we're cold. To me, there could be nothing better than looking out my kitchen window and seeing my life's work — the developed hill country, the recently planted forests and tidy, well-grazed paddocks. As I sip a cup of coffee at dawn, a mob of good ewes and prime lambs will string out along the laneway down a distant, razorback ridge like a dollop of thick cream. On a separate ridge road, the dust cloud from a 44-ton truck-and-trailer carrying logs recently harvested from our land will catch the morning sun, and the radio telephone will crackle with traffic as the staff engaged in the various spheres of Waipari's operations go about their business. It's very easy to feel proud and satisfied that the plan has finally come together.

One of the unintended benefits of living in the man cave is that it has meant that I have taken a step back from the epicentre of everyday station activity. It's a universal problem that as farmers get a bit older and less able to bounce back from the kinds of knocks that you inevitably collect wrestling a 400-kilogram cattle beast in the yards, we nevertheless struggle to take a backseat and let someone younger and more sprightly take over the everyday stock work. But I have come to accept that I am only worth $25 per hour doing farm work, whereas I'm worth $250 per hour in the office. Waipari is very lucky to have Nigel Hanan, yet another in a string of great managers. Nigel and his wife, Carla, run and care for the place as though it were their own. We have a great job demarcation line. If it can bleed, it's Nigel's department. If it can't, its mine. I tend to focus on admin, planning and growing the grass, while Nigel focuses on using it to best effect in growing the livestock.

***

Sometime after the man cave was completed, we finished major renovations on the homestead, with the aim of making it suitable for big family get-togethers and for occasionally renting out to selected corporate groups. It now sleeps 12, and seats a few more than that at a squeeze around a huge dining table my great friend George Wood and his sons made out of an 800-year-old totara tree that fell down in one of the forestry blocks. His brief was that it must comfortably seat 12 people, be strong enough for six girls to dance on (all at once!) and be able to shelter all 12 in a 6.5 earthquake. A major benefit of all we've done with the Waipari homestead is that it has become party central for the young — and it's lucky for some, perhaps, that the walls don't talk!

Speaking of the young, a perennial problem for farmers and farming families is succession. How, that is, is the value of a farming operation passed from one generation to the next in a way that is fair and equitable to all potential heirs? Not all farmers' offspring are keen on following them onto the land. Out of my lot, it looks as though Henry is the one most likely to follow in my footsteps. So how do you compensate those who want to pursue a different career?

The answer for Waipari and for me lies in the trees. The forestry side of things is run and managed as a separate business, and will provide an interest in the property and a cashflow that will ensure a smooth succession. Sorted — sort of!

In May 2014, with my mother, Lee, and 17-year-old Henry, I attended the Century Farm Awards ceremony in Lawrence in Central Otago, to receive a handsome bronze plaque in honour of the achievement of keeping the property in family hands for 131 years. It was a good night. In one of the speeches, a member of the organising committee commented on the diverse range of challenges I had faced and the solutions I had come up with to deal with them in order keep the business viable.

When we went up to receive the award and make our acceptance speeches, in his speech Henry told the audience that he was just counting the years until he could kick the old man off the farm and step up to claim the 150-year award.

The following day, with Henry taking very gentle care of his granny, we talked about what we would do with the time we had to kill in one of my favourite parts of the world. Henry dropped into the conversation that there was fresh snow in the Mackenzie Country. The Lindis and Burkes Passes were closed, but as Henry was driving Granny's 4WD Subaru Impreza, perhaps Granny might like to be driven home via the scenic route?

'It would be fun to drive in the snow!' he said.

So that's what we did. It was a long trip (nine hours, 450 kilometres) over hills and winding through snowbound valleys. I was mostly in the back seat, and with lots of time to reflect. I think it all began to sink in for the first time. Once I had dropped Henry safely back to his boarding house at College, I retired to the flat in Merivale belonging to my Aunt Nicky and Uncle Paul. I poured myself a large whisky and sank down in one of Granddad's old armchairs.

At that moment it fully hit home.

I had done it.

I had saved the family business, and steered it through to the point where the next generation was keen to take up the challenge. It was perhaps sad that neither my grandfather nor my father ever got to see the result of my life's work while they were still alive, but their ashes have all been scattered at the top of our highest hill, Papahope, to make sure they have a bird's-eye view of the progress. In fact, Granddad's were flown on with a load of super from the topdressing plane, and Dad's were a bit more accurately sprinkled from a hovering chopper. (My father had been very frustrated with my inability to be good at maths at school and had passed comment that without maths I would never be able fly a helicopter, a childhood ambition. When I turned 40, I began trying to achieve that ambition. It was my aim to be able to fly my father around the station to prove him wrong. However, as my skills extended only to flying a two-seater chopper with an instructor, there wasn't room to take another passenger. After his funeral, with Dad now reduced to a small box, it was all of a sudden very achievable!)

It was very poignant to be able to climb to the top of Papahope and feel that I was sharing my thoughts with them both, as we shared the vista of the station. Thirty years of worry, stress, frustration and fear, along with the challenge of facing yet another cold, wet, muddy winter morning, were now balanced with satisfaction and a quiet elation that I had completed the mission. I had thrown my heart to the top of a seemingly impossible hill-climb, and had reached the summit. At last.

And suddenly the big gulp of neat Famous Grouse backfired a bit. My throat burned, and like a frightened little boy, I burst into tears.

DON'T YOU LOVE IT WHEN A PLAN COMES TOGETHER?

Soon after my return from the Century Awards, in the middle of the winter of 2014, I sat down to honour my promise to John Falloon to write a book — this book. I faced the keyboard at my work desk that sits in one corner of the main room of the man cave, adjacent to my Grandfather Alwyn's 7-foot-long couch on which he told me he used to lie to dream up his episcopal sermons.

That's where I am today, more than three years later, putting the finishing touches to it. During the days I have spent working on the manuscript, I have occasionally looked out to where the sheep were grazing contentedly, or giving birth to triplet lambs, or (in the case of the lambs) gambolling on the hill-top in their first spring. At some point, as southerly storms have battered my little house and a fire has blazed in the wood-burner, each of those little lambs has been fine-tuned on winter forage crops in readiness for their departure on a one-way overseas trip. Then, by the wizardry of modern technology, their kill-sheets have arrived in my email in-tray just a stone's throw from where they were born, with every detail of their carcass weight, fat depth and wool weight, as well as their contribution to reducing the overdraft.

When I lift my eyes from the computer screen, my view is out across the valley to a montage of colour — 50 shades of green, if you like — where pine-tree blocks at various stages of growth, soil conservation plantings of willow and poplar, aged blocks of macrocarpa and eucalypt, Tasmanian blackwood together with the odd stand of kanuka/manuka add their distinctive hues to the patchwork of grass and forage crop, which themselves are shaded according to how hard they have been grazed, the fertility of each particular block, and whether it has been cultivated or is still in its natural state. The view from here would have been very different 100, 50 and even 30 years before.

The pasture is dotted with sheep or, in the case of some paddocks, Angus Pure beef cattle. And in summer, the dry forestry roads are wreathed with dust as heavily laden logging trucks rumble down the hill, often struggling to maintain braking traction on the steep track. Two radio telephones, one tuned to the forestry operation and one to the pastoral side of things, rattle with a stream of chat, sometimes with dire warnings of impending incidents, sometimes just with teasing banter as, say, when one truck manages to get out of a slippery skid site without a push from the D6 caterpillar dozer while his mate's 44-ton rig sinks in the sticky clay.

The stockmen riding their powerful 4WD bikes discuss over their two-way radios stock that may have eluded the muster on another shepherd's beat. Well, he or she isn't going to accept it without arguing the point, are they? It keeps me connected, as well as entertained.

This kind of technology wasn't dreamed of, let alone invented, 33 years ago when I first took over Waipari. And just now, the twin-engine, purpose-built rescue helicopter has passed overhead, following a set of GPS co-ordinates to pick up some poor bugger, possibly injured from rolling their four-wheeler. I can check immediately whose farm the mishap has befallen by logging into the helicopter GPS beacon that indicates where it has been and its last known position. Thirty years ago, that was the stuff of science fiction.

Soon after I took over the farm, I reluctantly gave away 4WD rallying. I'd just won a couple of national titles, so when people asked why I had quit when I did, I was able to truthfully say that it was a good time to stop, when I was on top. But of course such is the nature of the terrain at Waipari that I had plenty of opportunity to put my skills and passion for wrangling wheeled vehicles over and through obstacles to good work. Hillseekers and Mazda gave me plenty more. When mid-life crisis overtook me, I briefly considered buying a topless Mustang, or something like that, to keep me entertained. But I soon worked out that all it would do is collect speeding tickets. So instead I bought a Can Am Maverick, a radical, specialised off-road vehicle with which to trip around on the farm. It's suspension is so smooth that I can travel quickly without hurting my back.

Occasionally, I'm joined by like-minded others. We have a local group, the Bay City Sliders, who meet on many a muddy morning to get loose on the hills, and the first to roll has to wear a Bay City Rollers tartan skirt. We've had some interesting characters join us. Rod Drury, millionaire CEO of Xero was one. He returned from a run on one particularly steep track out the back looking pretty shaky. 'Come on,' I said, 'you can't tell me that's scarier than fronting up to a hall full of shareholders to report another loss?' He was non-committal, but, while he hasn't been back to tackle that particular track again, he has stood up at plenty of AGMs since!

As I write this on my laptop, sending each chapter instantly by email, via a cellphone running a high-speed broadband internet connection to a back-up somewhere in the Cloud, I can't help but wonder what the place will look like in another 30 years. That kind of speculation can be bitter-sweet, as you contemplate your life's work and become aware that it is all finite and drawing to an end. It won't be me who gazes with satisfaction over what's been achieved. I don't need my aching body to remind me I am on the downhill. Over the years I've been no stranger to gravity, and I'm very familiar with its pull. And as Henry has promised, when the time comes, they'll carry my coffin away perched on the back of Cardiac Arrest, my jetboat, towed behind the back of my 1953 Series 1 Land Rover 'Jock', to the sound of Fred Dagg singing 'If it weren't for your gumboots, where would you be?'

But looking out another window, my gaze falls on the sparkling blue water of the Pacific beating on the beach below me, and my thoughts turn to the next venture I've been cooking up. I'm thinking we might turn our most seaward paddocks to our profit and give commercial crayfishing a go. I've done some research, and it's an exciting prospect. That would have the three New Zealand primary production bases of farming, forestry and fishing all covered off.

So, gravity and all, look out! Because I still have some grip on the hill.

APPENDIX

THE ULTIMATE VEGETARIAN LAMB

My editor thought the following material might be a bit 'heavy-going' for your average reader not interested in the progress of farming lean meat over the years. But I thought it might be useful to sheep farmers, so they let me sneak it in at the end of the book! Here goes ...

Over time, Lean Meats has become a highly sophisticated operation, the very model of the kind of process that the meat industry calls for in terms of managing the whole 'gate-to-plate' supply chain. In fact, Lean Meats won a New Zealand Trade and Enterprise export award in 2003, and for a few hours in July 2016 Waipari was the only farm in the world to hold a GAP 4 (Global Animal Partnership) Certificate of Compliance for animal welfare standards in raising meat sheep, which was awarded by AUS-MEAT. Steve and Helen Knight down the road got their certificate a little later the same night, and many others have since been certified to comply as well. It gives you a significant marketing edge.

Lean Meats contracts our lambs a year in advance, which gives both producer (us) and processor certainty. It gives us a lamb supply plan to work to, stipulating targets for various mobs to achieve, and an indication of the daily live weight gain required to match those targets. Along the way, we give an update as to how the season is going. If we're having a difficult season and don't think we can provide the number of correctly weighed and specified lambs on the date we promised, we have the option of swapping our contracted space with farmers in another province who are also in the programme and who may be having a better season.

They kill our lambs at Progressive Meats, which serves as a 'toll processor', charging a fee per head for killing, chilling and (in recent years) performing some cutting and boning of the meat. This is then supplied to Lean Meats' American subsidiary, Atkins Ranch, which markets the meat through the Whole Foods Market chain, their main client. As a member of the Atkins Ranch Producer Group (ARPG), we get paid 90 per cent of the expected proceeds six weeks after kill (that is, when Whole Foods pays Lean Meats), and we then divvy up the remaining 10 per cent about six months later, by which time any unexpected extra costs or profits have been taken into account. If Lean Meats has had a very successful year, we will get a further payout based on realised profits. This has been as high as $20 per lamb in some years, but there have been quite a few years when there has been very little or nothing at all. It's been a very worthwhile relationship, hopefully for both parties. The meat industry catch-cry is that producers should have a very close relationship with their processors, and that's what Lean Meats has offered us from the outset. Thanks to our stake in the company, we have an interest in the whole supply process — we basically still own the lamb right through to the point where it is put on the supermarket shelf, and we even have a share in the Chevy van that carries it there, as well as the company's real estate in the States!

To further understand — or, let's face it, to test the conspiracy theory that the supermarkets take the lion's share of the profits in the supply chain — I bought a handful of shares in Whole Foods. Interestingly, the dividend from those shares didn't cover their opportunity cost (that is, the interest I would have received if I'd invested that money in other directions). But it encouraged me to take a close interest in the whole value chain, and this has subsequently informed my farming operation. From time to time, I've hosted the main meat buyers from the company on Waipari, bundling them into the Nissan Patrol to give them a tour of the farm on the bumpy, slippery tracks. The aim of the exercise is to show them that they can have confidence when they're fronting our meat to their customers, claiming as we do that it comes from 'perfectly happy lambs, contentedly eating salt-encrusted, pure green herbs and grasses, drinking the finest mineral water, while enjoying an uninterrupted sea view'.

They always seem impressed. But on one occasion, when I stopped and picked some clean grass and chicory leaves and doled out handfuls to my passengers along with an invitation to taste the freshness, I got a rap over the knuckles from head office. Apparently the word was that 'bloody Mark Warren was feeding meat industry VIPs grass when he should have been feeding them meat!'

In 2010, I got the chance to front our product to Whole Foods' customers myself when I was invited to travel to the States to be part of a lamb promotion team, cooking lamb samples at various Whole Foods supermarkets. We started in San Francisco, where we toured the Atkins Ranch meat-distribution centre, and then flew up to Chicago to work our culinary magic. At a university town somewhere out of St Pauls, Minnesota (I forget its name: we did a different supermarket each day, so we got a bit lost at times), I had the honour of serving a sample to Kofi Annan, former Secretary-General of the United Nations. He impressed me with his dignity and his mana. He was so measured in everything he did, right down to the way he held the box of spicy lamb sausages reverently as he asked carefully considered questions about the lambs and listened intently to my answers.

My modus operandi at these demonstrations was to set up my cooking table (we had a trolley table with frying pan and utensils) so that it pretty well blocked the thoroughfare. Everyone had to line up to get past. I figured that people would probably accept a sample if they were detained right there, and that if people saw a queue and lots of interested people milling around they would join in as well. We took care to festoon the vicinity with photographs of the lambs out on the coast, munching on lush, green grass, and of the boys picturesquely at work on the farm. That guaranteed lots of questions about New Zealand and New Zealand farming. The more obviously educated clientele seemed to have visited New Zealand already, and pretty much all of them declared their intention to return some day.

We would stand there wreathed in the irresistible aromas of seared lamb, offering tastings and waxing lyrical about the 'clean and green' aspect of our property. I must have cooked about 3000 samples over the course of the two weeks, and by the end I had it all down pat.

'Try some vegetarian lamb, ma'am?' I would say.

'What's that?' she would reply, looking confused. 'Is it made out of soy or something?'

'No, it's just very special, grass-fed lamb — very different to the grain-fed lamb you've probably tried before.'

She would try it, chewing thoughtfully.

'Goes down a treat with a nice merlot from the Hawke's Bay in New Zealand,' I would assure her. 'And it's best cooked in olive oil from The Village Press, not far from my farm. You'll find that one aisle over.'

'So you're the rancher?' she might ask.

'Yes, ma'am,' I would say, gesturing at the photos with a show of pride. 'These are all my lambs. They're more than that. They're my girlfriends! Miraaaaanda. Baaaaaabara. Maaaaarian.'

'Oh, no! I'm eating your cute little girlfriends?'

'Don't worry,' I would wink. 'We only eat the bad-assed little ram lambs.'

'Phew. That's all right, then. So what do you do to make your cute little lamb buddies feel special?'

'Well, ma'am. Every one of my girls gets a Brazilian at the beginning of summer ...'

Most of those I engaged in conversation seemed inspired enough by my caring pastoral regime to pop a box of lamb patties or spicy sausages into their trolley. If they looked a bit indecisive, I would place a box of frozen patties in their trolley, or volunteer to run over and 'grab a perfect rack' for them from the full-serve butcher. The real challenge was to get them to buy a pound of 'lamb rack'. At about $80 per kilogram in New Zealand terms, it was the second most expensive meat in the supermarket.

The managers of the meat departments of several of the supermarkets raved over the effectiveness of my salesmanship. 'You just sold a week's worth of lamb in a four-hour demo session,' one said, shaking his head in wonder. Their enthusiasm cooled as staff began finding half-thawed packets of 'spicy lamb sausages' and lamb racks leaking blood hidden behind packets of cornflakes two aisles further on ...

Out of all those tastings, I managed to convert two vegetarians, but I did have one spitter. One or two very glamorous housewives — bored or desperate — invited me home to cook the lamb for them in person. These invitations were fun to entertain for a moment or two, but the possibility that I might be AWOL the next day and let the team down prevented me. Perhaps it was wishful thinking, anyway, although one woman even sighed and asked whether she could come and live on my ranch with me. By the look of her, I didn't think hard work on the farm was at the forefront of her mind.

It was a most interesting time and helped me understand the supermarket business from the other side. I finished up with a very healthy respect for supermarket managers, and I was very glad to go back to farming at the end of it. A while later, Forbes Elworthy (from an old and famous New Zealand farming family) was leading a crusade to get more value back to farmers for their lambs. When as part of his presentation he showed a model indicating the various cuts of a lamb and their retail values, in order to argue (quite rightly) that we weren't getting our fair share, I stood up in front of the hall-full of 400 other farmers and challenged his figures. I pointed out that not every cut sold for the full face value it carried in the chiller. There was old, nearly expired stock that had to be made into curries at a reduced price, and the odd reject that dropped on the butcher's floor as well as dissatisfied customers who complained and got their money back.

Forbes rather pompously challenged my facts, and asked where I had got my information.

'I've recently spent two weeks in the Whole Foods supermarket chain in the States,' I told him. 'I spent quite of lot of that time behind the counter in the butchery section, so I saw what goes on at first hand.'

I supported what Forbes was trying to do, but he soon abandoned that crusade and moved on to another. It was interesting to see him following in the footsteps of his father, Sir Peter Elworthy, some 25 years later doing a variety of things, including co-founding Craigmore Sustainables, specialist managers of farms and forests.

Initially, we aimed to produce lambs for Lean Meats of 45 kilograms live weight, which killed out at an average of 20 kilograms per carcass. Over time, we have increased the average live weight to over 50 kilograms, to produce a 23.5-kilogram lamb. At one point there was an opportunity to supply an even heavier carcass of up to 30 kilograms. In October 2011, we killed a line of lambs that netted $7.70 per kilogram, and averaged over $201 per head. That couldn't last — in fact, it was uneconomic at the time, as the meat companies ended up paying far more for lambs than they could get for them. Inevitably, the market corrected in subsequent years.

Nevertheless, and largely thanks to initiatives such as Lean Meats and Whole Foods, the lamb price has risen threefold in the period I have been farming. To put it in relative terms, you can plot how many lambs it has taken to buy a car, ute or tractor over time. On this analysis, the results are encouraging. In 1973, the schedule for a benchmark 13-kilogram premium lamb was $7. A brand-new, 6-cylinder Holden Kingswood station wagon, or a short-wheel-base, canvas-top Land Rover, or a 50-hp Ford 5000 two-wheel-drive tractor all cost about $4000, or about 570 lambs. Nowadays, the average lamb — about 18 kilograms — will fetch about $100. For 580 lambs, you could buy a station-wagon Land Rover 90 Defender. For 440, you could buy a well-appointed 4WD double-cab diesel ute, or a Holden Commodore wagon for about 430. A 60hp Ford New Holland tractor would cost a mere 340 lambs.

The sharp-eyed amongst my readership might have noticed that the comparison is between a 13-kilogram lamb in 1973 and an 18-kilogram lamb today. That's indicative of a market trend. Larger lambs are the norm: 18 kilograms is the new 13 kilograms.

If I look back at total lamb production on Waipari, we could usually expect to achieve 85 per cent lambing in 1973. With 6000 ewes, and allowing 25 per cent (1500) kept as replacements rather than sold, in theory we had 3600 lambs to sell, counting for stock losses. If we sold these at $7 per head, our lamb crop was worth $25,200. At $4000, a Holden Kingswood station wagon therefore cost 16 per cent of the lamb crop. Today, we would expect to average 135 per cent from 4000 ewes, giving us 5400 lambs, which, allowing about 25 per cent (1000) for replacements, works out at 4400 lambs to sell for around $440,000. The Holden Commodore mentioned above costs about 10 per cent of the lamb crop, and it is a much more advanced and refined product than a 1973 Kingswood, having almost twice the power and using about half the fuel.

So what does all this tell us? It tells us that over this period the lamb price rose by a factor of about three. Costs, meanwhile, hardly stayed static. It's always a balancing act. And as I recently stated in an interview, for sheep farming to hold its own against other land-use options we need to be able to buy a ton of superphosphate (costing $316) with the proceeds of two good winter lambs. We are just under that at present.

⁂

With lamb prices relatively buoyant and wool prices static, it's little wonder that farmers focus on lamb production. Any extra wool weight that can be bred into an animal is mostly pure profit. The business of being a 'fat lamb farmer' of the early 1980s and 'a lean lamb finisher' of today is very different. High-performance lamb production has evolved into a science itself, with an emphasis on qualities such as worm resilience, higher carcass yield, better carcass conformation (that is, growing meat on the parts of the animal that are worth more) and improved leanness. There has even been a concerted effort to produce lambs that don't need dagging.

Mother Nature plays a huge part in it all, and if properly enlisted in the partnership she can have a profound effect on your bottom line.

One of my early ambitions for Waipari was to source a hardier and more self-sufficient Romney breed than we had been using. The original rams had been bred on a flat-land farm with an emphasis on size, not efficient performance. For many years my good friend and fellow Hawke's Bay farmer Bay de Lautour had been selecting on hardness and self-sufficiency as well as easy-care lambing. He and his son Hamish were aiming to breed a sheep that would deliver 150 per cent lambing without being shepherded at lambing time, and for the sheep to have a natural resilience to parasites. I started using his Te Whangai breed of Romney, and, while it took 21 years, I managed to achieve 150 per cent lambing. That's a sight better than the 56 per cent Waipari managed in 1983. Bay has been occasionally heard to boast that his breed has tripled some clients' lambing percentages!

Our sheep are now very low-maintenance. These days, we put the ewes into sheltered paddocks and go on holiday at lambing time. Unless there's a big storm, we expect a paddock tally normally in excess of 150 per cent, and with about a 2 per cent loss rate in the ewes. Under the old intensively shepherded system, the death rate could be up to three times that. When I was interviewed recently about farming practice and the Te Whangai breed in particular, I was quoted as saying 'the best way to lamb a Te Whangai-bred Romney was from a chairlift'.

In recent years, Bay and Hamish have taken the use of Mother Nature to another level. They have been refraining from drenching their stud flock, which has the effect of placing huge selection pressure on lambs, finding out which lambs can thrive even with a large worm burden. This has evolved a sheep that no longer needs drenching, or that needs to be only minimally drenched under certain conditions, with a corresponding reduction in production costs and an increased profit — making for a better lifestyle for both sheep and farmer!

Technology has helped enormously. Once, drafting the killers from the underweight lambs was a labour-intensive job requiring a quick eye and a strong hand on the drafting gate at the end of the race, which you flicked to and fro to direct each animal into one pen or the other. Now, in order to match our animals to our stock-raising plan, we use an electronic stock-weighing system that incorporates a padded, compressed air crush (a kind of clamp to hold the animal in position) to catch the lamb, weigh it and to decide which pre-programmed weight-range mob it is to be sent to. Once weighed, the lamb is released into the correct mob at the push of a button that opens an air-operated drafting gate. A good man and even better dog can weigh up to 500 lambs an hour.

If we have contracted 200 lambs to be delivered on a certain date, about 250 lambs will be re-weighed a week before that date, to confirm that we will have at least 85 per cent of the contracted mob up to target weight and ready to be processed on the day required.

Typically, that mob would also be redrafted in up to five ways. One mob would be for animals that are too light. Another would be for lambs that meet the killing specifications, but which have room to improve. The third mob will be for lambs that are within 5 per cent of the target weight, and which we will often put on very high-quality 'rocket fuel' — finishing feed, most lately forage green-feed barley — for a week so that they gain up to 600 grams a day to reach the perfect target weight. The fourth mob, those animals that are bang on target, are fed maintenance, while the overweight lambs (or those which exceed the stipulated measurement of fat over the seventh rib) will be put in the 'Jenny Craig paddock' where they will be on a restricted diet so that they can shape up.

A week before the processing date, we confirm by email the number of lambs we need to book killing space for, and the processor in turn emails the trucking company a list of the various mobs in the district so that they can plan an efficient pick-up itinerary.

On the day of trucking, the lambs are re-weighed to select those closest to the perfect target live weight. If we manage to ship 85 per cent or more of the number we have contracted to supply at the target weight (which is set up to a year before), we receive an 'in-full and on-time' extra payment. Similarly, we also get a self-drafting fee of 85 cents per head; and an Atkins Ranch Producer Group premium of 10 cents per kilogram (as a reward for being able to prove we have met their requirement that the lambs are bred on the farm they are finished on, that they have received no antibiotics, that they are only grass-fed); and a quality-assurance-approved payment of 50 cents per head for having achieved and maintained a strict animal welfare code of practice. All in all, the extra, above-average requirements are worth around $4.20 per lamb and amount to approximately $15,000 per year. Given that this is roughly equivalent to the amount I draw for personal spending every year, it's a real incentive to keep our standards up.

Within 12 hours of the lambs being processed, we are emailed a kill-sheet with the actual carcass weights and GR measurements (the measurement of the layer of fat around the seventh rib), so we can confirm our live-weight/carcass-weight yield factor calculation. Since the yield can vary from 38 per cent to 51 per cent, this is an important number, because we draft on live weight but are paid on carcass weight. Some skill and experience is required to get it right. We have been greatly assisted by the recent inclusion of numbers for the average live weight at the works, which, taken together with the average carcass weight, gives us an average dress-out percentage. There is always a difference between live weights at the farm gate and the live weight at the works, sometimes up to 5 per cent if the lambs hang around waiting too long. Naturally, we work very hard to minimise the time between the lambs leaving the farm and being processed, with a target time of four hours. A day's wait can be a whole kilogram of lost live weight — five whole bucks!

An added advantage of this quick feedback is that someone who is learning the art of drafting the perfect lamb can receive an indication the very next day of their skill level. They can earn a bonus for themselves if they hit the jackpot. It's always a pleasure to be able to congratulate a member of our team if they achieve the magic number and get 100 per cent accuracy in their drafting. Our self-drafters seldom score below 95 per cent, but if they do, we can move quickly to find out why.

In the good old days, England would take any lamb that was fat enough, and lambs were shipped as a whole frozen carcass. Now, the lamb carcass is carefully dissected by highly trained butchers at the Progressive Meats plant, each of them being rewarded according to the amount of high-quality meat they can recover from each carcass. The various, specific cuts of chilled meat are then shipped in special, gas-flushed packaging to whatever international client is paying the most for that cut. The rack (glorified chops, really), for example, might go to Whole Foods in the United States, the leg to Europe, and parts of the shoulder, bones and offal to China. The woolly pelt, if supplied to specification, may be used to make high-fashion ugg boots or fluffy baby rugs, and the cleaned intestine tubes (runners) are used to make casings for high-quality, gourmet sausages. It's the source of considerable pride to all of us to be involved in the value chain of one of the world's premium meat products. And while the rest of the world seems to be chasing lower commodity prices, the market is rewarding us for our efforts. The lamb price has been steadily tracking upwards for some time now.

Our medium-term goal is to produce 4000 24-kilogram lambs that fit the Atkins Ranch Producer Group specification. At current prices, that would give a return that would cover the average cost of capital, and provide a small profit for our efforts, with, of course, the added incentive of solid, long-term capital gain. Interestingly, our current gross income per year is about the same as what the whole business was worth about 28 years ago. Farming has proven to be a worthwhile investment, for all the times when I could have walked off the property for a fixed salary in a warm workshop somewhere.

PH (PERSONAL HEALTH) LEVEL IN THE TOP PADDOCK

Recently we lost a good mate. One of the best. A man who seemed to have reached the pinnacle of personal success, with an incredible wife and children. He was a top achiever, a winner of the Hawke's Bay Farmer of the Year award, a top nationally recognised rugby player and a highly respected sportsman. At a recent Farmer of the Year field day, his farm looked as perfect as could be dreamed. But the pressure to keep it at that level was huge. And the season seemed to turn against him. Sadly the end result was a massive funeral, held in the Hawke's Bay A&P Exhibition Hall. The venue was overflowing, with well over 1000 family and friends in attendance. It made TV1's *News at 6*. A mighty totara had fallen.

The blokey Kiwi farmer stereotype is a celebrated image. We're meant to be big, tough, hard hill-country men. Words aren't meant to be wasted discussing feelings. It's no surprise, then, that many of us blokes don't feel comfortable admitting weakness — that things have got on top of us. We are conditioned not to show any form of weakness. Not to give an inch. That's celebrated, admired and worshipped in our culture. But what we need to remember is that even the strongest trees bend in a storm.

Often life is not as perfect as it might look to others ...

Recently, I snapped. The season had been tough, we were selling stock for less than we had hoped. Four years of very dry conditions had been followed by a very wet winter. Whenever you tried to get a job done on the tractor, you were in constant danger of sliding off a hill or getting stuck in mud up to its axles. Feed was short, and farmers hate to see their stock go hungry. Due to very low wool prices, the overdraft was rapidly climbing towards its limit, and I had been threatened with a mortgagee sale by the lawyer acting for the family I felt I had supported very well over the past 33 years. Spare cash was tight. I was relying on my off-farm 4WD training income and savings to live on. Farming is a hell of job sometimes, made harder every year by the cost of compliance.

When I snapped, it was my dear, darling Julie who felt the impact. The outcome was that Julie booked me in to see my GP, Dr Bob. And I was given information I didn't like: as Dr Bob put it, I tended to maintain my tractor much better than I maintained myself. As I do with my business planning, when I receive advice I don't agree with, I ask three other people for their opinion. If they agree with the advice, I accept it. The outcome of his advice was that I would occasionally take a tiny wee pill when I was getting stressed, tired and frustrated, which would allow me to sleep through the night, and let not only my body rest but my brain also. It's worked wonders.

Everyone is different, but you should never be afraid to (a) ask your doctor and Rural Support for help, (b) take prescribed medicine, if that's what the doctor's advice is, and (c) have a break off the farm for a few days to recharge your batteries.

The only other advice I would give to someone under pressure is to have a mentor — someone you trust, and someone who understands your business and is successful in their own right. And someone you feel comfortable discussing all your issues with. We all need to do our bit to reduce the rural suicide epidemic. Often a problem shared is a problem halved. If you know someone who is struggling to make things work, make sure you reach out to them — even if they appear to be fine on the outside.

ACKNOWLEDGEMENTS (AND APOLOGIES)

A book like this doesn't happen by chance — it's a product of many people's input.

Not necessarily in order of importance, but more a timeline, thank you to:

- Peter Smart, my English master at College, who must have taught me some writing skills, and didn't seem to regard me as dumb, unlike some.
- My accountant, Pita Alexander, who has been more than just a number-cruncher, but also a mentor and wise adviser, guiding me through many, sometimes violent, business storms.
- My mother and father, who allowed me the freedom to go my own way.
- John and Phillipa Falloon, who urged me to write a book about my life's experiences.

- Julie Holden, who has supported me through many troubled times, and also used her English skills to make readable my dyslexic writing.
- John McCrystal, who tidied up my first draft into a readable order.
- Thank you to the HarperCollins team for educating me in the ways of publishing, not least Sandra Noakes who attempted to Aucklandise me at the publisher's meeting with a beetroot latte!
- And last, but by no means least, Alex Hedley, for persevering and transforming my stories into (hopefully) an entertaining book.

Finally, also a big thanks to all those who have worked on Waipari and those who supplied us with their services to make it what it is today. And apologies to those who are still nursing sore toes from having them stood on at times!

Almost another furrow in my brow (page 158)